The Shadow of the Black Hole

The Shadow of the Black Hole

JOHN W. MOFFAT

OXFORD
UNIVERSITY PRESS

OXFORD
UNIVERSITY PRESS

Oxford University Press is a department of the University of Oxford. It furthers
the University's objective of excellence in research, scholarship, and education
by publishing worldwide. Oxford is a registered trade mark of Oxford University
Press in the UK and certain other countries.

Published in the United States of America by Oxford University Press
198 Madison Avenue, New York, NY 10016, United States of America.

© John Moffat 2020

Library of Congress Cataloging-in-Publication Data
Names: Moffat, John W., author.
Title: The shadow of the black hole / John W. Moffat.
Description: New York, NY : Oxford University Press, [2020] |
Includes bibliographical references and index.
Identifiers: LCCN 2019048687 (print) | LCCN 2019048688 (ebook) |
ISBN 9780190650728 (hardback) | ISBN 9780190650742 (epub)
Subjects: LCSH: Black holes (Astronomy) | Gravitational waves.
Classification: LCC QB843.B55 M64 2020 (print) | LCC QB843.B55 (ebook) |
DDC 523.8/875—dc23
LC record available at https://lccn.loc.gov/2019048687
LC ebook record available at https://lccn.loc.gov/2019048688

1 3 5 7 9 8 6 4 2

Printed by Sheridan Books, Inc., United States of America

To my wife, Patricia

CONTENTS

ACKNOWLEDGMENTS

My greatest indebtedness is to my wife, Patricia, whose patience, encouragement, and editing skills made this book possible. I thank my editor at Oxford University Press, Jeremy Lewis, for his enthusiasm about this book project and his patience with the delivery of the final draft while we waited for necessary data to come in.

My thanks go to Michael Landry, director of the LIGO observatory at Hanford, Washington, for his hospitality during our visit and his valuable help in getting the details right about gravitational waves and the LIGO observatory. My sincere thanks go to my collaborators Martin Green and Viktor Toth for their insights into the physics and astronomy in this book, and for their careful reading of the draft. I appreciate the support that the Perimeter Institute for Theoretical Physics has provided me during the writing of this book.

Prologue: LIGO

We left the city of Richland, Washington, on the desert road leading to the Hanford LIGO site to spend two days visiting the now-famous gravitational wave detector observatory. When my wife, Patricia, and I left our hotel, the sky was overcast with dark-gray clouds. There was a chilly, damp wind, and we had taken our coats to shield us from a typical March day in eastern Washington. The road to the site was straight, and we looked out over the unobstructed landscape covered with sagebrush and tumbleweeds blowing in the wind, all ringed by purple mountains in the distance. Hanford was one of the sites in the Manhattan Project, which produced the nuclear bombs that destroyed Hiroshima and Nagasaki in August 1945. We met very little traffic on our way, other than trucks leaving the infamous reactor site left by the Manhattan Project. The legacy of that project, the largest nuclear waste site in the United States, has contaminated the groundwater underneath 61 square miles of the site, and it threatens the headwaters of the Columbia River.

At a road crossing, a white-and-red-striped security bar prevented us from entering the old reactor site, and we turned left toward LIGO. We had an appointment to meet the head of the observatory, Michael Landry, at 9:00 a.m. No prominent signs told us that we had arrived at the LIGO site; however, off the road to our right, a large dark-blue tower rose up between white and Mediterranean blue buildings with the letters LIGO painted on its side, which stands for "laser interferometer gravitational-wave observatory." This site in Washington is one of two in the United

States that work in tandem to catch gravitational waves originating out in space. The other site is in Livingston, Louisiana. I wondered whether the buildings were alike at both sites.

We approached a building with a sign saying VISITORS and parked the car. I shivered as we stood for a few moments viewing the site, which seemed desolate and empty of human beings. Perhaps it had this abandoned feeling because it was still early in the morning. Patricia and I entered the visitors' building and found ourselves in a large, also-deserted lobby, with displays of the LIGO project on view. A secretary came into the room and greeted us, and we explained we had an appointment with Dr. Landry to visit the laboratory.

At that moment, Mike Landry appeared and greeted us warmly. He was tall and athletic looking, with an infectious smile. He led us into his office at the back of the building and invited us to take seats near a blackboard covered with mathematical formulas. He offered us coffee or tea. He explained how our visit was planned for the day, and didn't waste much time before getting into a detailed description of the observatory setup. As he spoke, my eye wandered to the blackboard, and there, at the bottom right, was written in chalk "170104." It immediately occurred to me that this could possibly indicate the date for a fourth gravitational wave detection from the collision of two inspiraling black holes far away in space. The new LIGO run had started in November 2016 and surely, hopefully, they had detected another black hole collision by now, in March 2017. Two strong signals of gravitational waves had been discovered already—on September 14, 2015 (GW150914), and on December 26, 2015 (GW151226)—and a less convincing one on October 12, 2015 (LVT151012). The names of the events are simply the dates on which they occurred, with "GW" standing for "gravitational wave" and "LVT" standing for "LIGO-Virgo transient."

My eyes returned to Landry's face and he smiled because he realized I had picked up on the date of the fourth gravitational wave detection event. He chuckled and said, "Yes, that is the fourth event."

"Well," I said, "I wasn't going to mention anything about what I saw on the blackboard, because I'm sure the detection of the event is under a cloud of secrecy."

Mike chuckled again and said, "Indeed, whatever passes between us now regarding this event should be kept quiet until we officially announce the detection."

He continued with his description of some of the basic technical issues and problems faced by the LIGO team, and how they had resolved them successfully. At a pause, I could not resist the temptation to press him about the latest detection. "I presume the LIGO collaboration has already analyzed the data?"

"Yes," he said, "that's correct. The masses of the two merging black holes are 31 and 19 solar masses."

The gravitational waves detected by LIGO originate from the crashing together and merging of two massive, faraway compact objects—black holes or neutron stars. The LIGO team is able to figure out information such as the distance of the merging objects from Earth, their masses, and their spins by clever analysis of the gravitational wave data coming in.

"Aha!" I said. "So the primary black hole masses are still large, comparable to the primary black hole masses for the first event, GW150914. What about the spin alignment for the merger?"

"It's negative."

"Negative?" I burst out. "So the spins are significantly misaligned, like the other two significant events—GW150914 and GW151226. Very interesting!" One would expect that two black holes in a binary system would be spinning around each other like two ice skaters, with their axes similarly aligned. However, three gravitational wave detections had now shown the spins of the merging black holes, oddly, to be significantly misaligned with each other. An alternative interpretation of the data for the spins is that the merging black holes are spinning or rotating slowly.

"I presume this is just as strong an event as the first two, around five sigma?" I continued.

"Yes, it's a strong event, depending on how you interpret what is meant by the standard deviations."

Patricia had her notebook out and was scribbling shorthand notes of our conversation.

"Well," he said, "let's begin the tour of the laboratory. We have a maintenance shutdown between 8 a.m. and noon, so we have time to visit the control room and several other important parts of the LIGO setup."

We left the visitors' building and walked to the main part of the observatory. I made some offhand comments about the desert landscape and the wind that was sweeping tumbleweeds toward us. Mike said that visitors are often surprised by the desert landscape, expecting Washington state to be Pacific Northwest rainforest throughout. During our walk to the observatory, we were surprised to learn that Mike was originally Canadian.

In the main building, we made our way to the control room (Figure 0.1), a large room with banks of colorful computer monitors on the walls facing us. LIGO personnel sat around peering at their own desktop monitors and laptops. We were introduced to two young physicists who were in charge of analyzing the incoming data. Mike pointed to a large, central monitor on the wall showing jagged red, yellow, and blue curves. He said, "That's

Figure 0.1. LIGO control room. Credit: Patricia Moffat

the data analysis of the signal-to-noise ratio for incoming data. You can see several spikes on the plot that are caused by seismic shifts in the Earth."

"None of the monitors show an actual detection of a gravitational wave event?" I asked.

"No, no," Mike said, smiling, "you never see them. The data have to be analyzed by the LIGO collaboration. You never directly see any incoming gravitational wave event."

"When the detector is running, then you must have night and day shifts for computer and data control?" I asked.

"Yes, people are here around the clock. In fact, very early in the morning on September 14, 2015, I got an e-mail saying there had been a potential detection of gravitational waves. I was asked to come to the laboratory as soon as possible."

"The analyzed waveform for the first event was seemingly perfect," I said, remembering the dramatic data chart I had seen for the first event. "It looked like one of the numerical relativity computer templates. Could it have been a blind injection? I understand that you had a system set up to insert blind injections into both the Livingston and Hanford sites to sim- ulate a real gravitational wave detection to check whether the detection system was working."

Mike said, "Yes indeed! When I first saw the result, I have to admit I was suspicious and asked whether we were in a blind injection phase. Once the answer came back no, we moved to a phase where any injection could be vigorously investigated. Every effort was made to make sure there was no extraneous cause of the first event."

After a very interesting tour of the impressive LIGO detector, during its maintenance shutdown (see Chapter 7), we joined a special group of researchers in a large meeting room. Mike, surprisingly, had invited us to a lunch session in which key members of the collaboration were preparing a manuscript detailing the latest gravitational wave detection event for publication in *Physical Review Letters*. In Mike's instructions for our visit, he had suggested that we bring our own lunches along. There was no cafeteria providing food at the site, only a small coffee room with coffee, tea, soft drinks, and a refrigerator for the team's bagged lunches. Patricia and I seated ourselves against the wall of the meeting room while members of the editorial team filed in with their lunches. The head of the editorial team had prepared a basic draft of the paper in collaboration with Mike and others. The group proceeded painstakingly through the draft of the paper section by section, with many people making comments and suggestions. After the session, I marveled aloud at how they managed to prepare a manuscript with more than 1000 authors, which was roughly the size of the worldwide LIGO collaboration. Mike explained that the LIGO/Virgo collaboration has small writing teams for papers, and this one was at LIGO Hanford. The writing leader was Keita Kawabe.

We felt honored to be present at this historic moment. The detection of a third major event over the period of some 40 days of running time during the second LIGO run was important because it provided evidence beyond a doubt that gravitational waves had been detected emanating from the cataclysmic merging of two black holes 1.4 billion light years away.

Near the end of the day, after our tour of the LIGO site with Mike, the editorial meeting, and interviews with several of the senior members of the collaboration, as we were preparing to leave and return to our hotel in Richland, Mike asked me whether I might be willing to give a talk the next day to the collaboration and students. Mike knew I had recently been working on the LIGO data myself, in connection with my gravity theory, MOG, and had prepared a paper for publication on MOG's predictions for LIGO data. As it turned out, I had not brought my laptop along on this trip, but by sheer coincidence I had a memory stick in my trouser pocket containing a talk about my gravity theory I had given at the Perimeter Institute two days before our departure to Seattle. At the security check at Pearson airport in Toronto, while emptying my pockets of metal objects, I had fished out the memory stick. I had looked at my wife and said, "Why did I bring this along?"

She had laughed and said, "Well, you can always give a talk!"

I handed the red-and-black memory stick to Mike, who popped it into his laptop and pulled up the pdf of my talk.

The next morning, when we arrived at 9:30, Mike introduced us to Rick Savage, a senior member of the LIGO collaboration. He took us to his office, where we sat down and my eyes immediately locked on to a large poster of a painting by Mark Rothko, one of my favorite abstract painters. I turned to Rick and said, "I see you have an interest in abstract art."

Rick smiled and responded, "That's an old poster given to me by a friend years ago."

He then explained, with much enthusiasm and obvious expertise, how the laser optics instrumentation at the detector site had been developed and improved over many years. Perhaps the most serious technological challenge in the LIGO project is removing the considerable noise in the data that obscures true signals—noise such as earthquakes, rumbling nearby traffic, as well as thermal distortions in the test mass mirrors at the ends of the 4-kilometer-long detector arms, and even quantum noise in the laser. The apparatus must detect movement in the arms of an unbelievably small magnitude: equivalent to a length of one one-thousandth the diameter of a proton. Such is the extraordinary accuracy of this modern technology, which had been developed over decades of research by pioneers such as Rainer Weiss, Ron Drever, Stan Whitcomb, theorist Kip Thorne, and other important figures.

During Rick's intense and detailed description of the laser and optics instrumentation, I began looking at my watch nervously.

"Rick, I'm supposed to be giving my talk at 11:00 a.m. and it's now just after 11:00."

Just then, right on cue, Rick's phone rang and he said into it, "Yes, Mike. We're on our way."

In the large seminar room, many of the 40 or so members of the laboratory were present, waiting expectantly. I began my talk by thanking them for their gracious hospitality, and said how honored I was to be able to give a talk about gravity and gravitational waves at this historic time and place, with the amazing discoveries of gravitational waves and merging black holes now underway. My talk was mainly about the consequences of my modified gravity (MOG) theory and its predictions for gravitational waves and black holes. I described MOG's potentially significant deviations from

Figure 0.2. Mike Landry and John Moffat at the LIGO detector arm.
Credit: Patricia Moffat

Einstein's general relativity. I also mentioned the upcoming Event Horizon
Telescope (EHT) observations of the supermassive black hole at the center
of our Milky Way galaxy and the one in the center of the galaxy M87. This
project will be observing the shadows of the black holes cast against the
bright background of their accretion flow of gas. This is the other major
experiment about to begin running on black holes, the first observations
of which would commence the following month, in April 2017. My MOG

had somewhat different predictions for the results of both LIGO and the EHT than Einstein's general relativity.

During the question period after my talk, the audience showed a lot of interest. They were receptive to my alternative gravitational theory and to the LIGO data verifying or not verifying Einstein's general relativity for strong gravitational fields.

After we said our farewells, Mike took us on an impromptu tour of the outside of LIGO. We stood at the conjunction of the two long LIGO arms (Figure 0.2), marveling at the distance the laser beams of light were traveling inside them, and the delicate engineering we had learned about that made the detection of a gravitational wave possible.

As we began our drive over the mountains back to Seattle, I reflected on our visit and on what the science at LIGO means. The merging of two black holes into one larger black hole represents the strongest gravitational forces that have been observed anywhere to date. This is also the first time that black holes have actually been detected, or "heard," almost directly. The LIGO detections, in fact, provide the most dramatic evidence yet for the very existence of black holes. The LIGO work, along with the upcoming EHT observations, is opening up an exciting new era of astronomical observations that promise to change profoundly our understanding of the universe.

1

Gravitation and Black Holes

Black holes are one of the extraordinary phenomena in the universe whose existence was surmised not by observations, but by theory. The black hole is a prediction of Einstein's 1915–1916 gravitational theory, general relativity, which replaced Sir Isaac Newton's gravity theory, published in his famous treatise *Principia* in 1687.[1]

In 1784, Reverend John Michell, a Fellow of Queens' College and Professor of Geology at Cambridge University, had already envisioned what we now call black holes. He asked what would happen if a star's gravity were so strong that its escape velocity—the speed at which a rocket, for example, would have to travel to leave the star—exceeded the speed of light? Michell realized that any light emanating from the star would have to fall back to its surface. He speculated that the escape velocity would exceed the speed of light for a very massive star, making the star invisible to an observer.

Independently, in 1796, French astronomer and mathematician Pierre-Simon Laplace postulated the existence of "non-luminous bodies" on the basis of his similar work on escape velocities and massive bodies. He found in his calculations that if you made the Earth half its size but kept its mass the same, the escape velocity would double to 22 kilometers per second. If you kept squeezing the Earth smaller and smaller, eventually

1. A. Einstein, "Die Grundlage der allgemeinen Relativitätstheorie," *Annalen der Physik*, **49**, 769–822 (1916).

the escape velocity would exceed the speed of light. Laplace calculated that this would happen when the Earth reached a tiny diameter of only 1.8 millimeters. The concept of a black astrophysical body was born. Laplace published a mathematical proof of this idea in 1799. The proof took the form of an essay proposing the existence of invisible bodies (which we now call black holes).

The fact that Michell and Laplace came up with the same idea is a remarkable coincidence, as there was little scientific contact between England and France during those troubled times in French and British history. However, others have questioned this coincidence, claiming that Laplace did hear of Michell's theory that gravity could prevent light from emerging from a star, and he produced a mathematical proof out of curiosity. Laplace did not include this proof in the original edition of his book *Exposition du Système du Monde*, and he also excluded it from subsequent editions because he was skeptical of the actual existence of invisible stars.

EINSTEIN GRAVITY AND BLACK HOLES

More than 100 years later, when Einstein published his theory of general relativity, he predicted that gravity could bend light rays. Once you acknowledge that gravity can bend light, you realize that strongenough gravity can bend light in on itself. In 1916, Karl Schwarzschild (Figure 1.1) worked out that, for a given mass of an astrophysical body, there was a specific radius at which light would be unable to escape the body.[2] This became known as the *Schwarzschild radius*. The formula tells us the size that an object with a given mass would need to be for the escape velocity to equal the speed of light. For example, for the mass of the Sun, the Schwarzschild radius is about 3 kilometers—the radius at which the Sun would cease shining, to an observer, and would collapse under its own gravitational pull into a black hole. Clearly the Sun, with a radius of about 696,000 kilometers, is

---

2. K. Schwarzschild, "Über das Gravitationsfeld eines Massenpunktes nach der Einsteinschen Theorie," *Sitzber. Deut. Akad. Wiss. Berlin, Kl. Math.-Phys. Tech.*, 189–196 (1916).

Figure 1.1. Karl Schwarzschild. Photo credit: Wikimedia Commons (Berlin-Brandenburgische Akademie der Wissenschaften, Archive)

not yet a black hole. The Schwarzschild radius for the Earth is about 1 centimeter. It is interesting that Laplace's estimate for a dark Earth was close enough to the value now estimated from the Earth's measured mass and the measured speed of light. In addition to the Schwarzschild radius, the Schwarzschild solution had a singularity at the center, where the density of matter is infinite.

Karl Schwarzschild was in the German military on the Western Front in 1916 when he solved Einstein's field equations. The field equations constitute 10 partial differential equations determining the interaction of matter with curved spacetime. Schwarzschild developed a rare autoimmune disease, pemphigus, and was hospitalized in Frankfurt, where he was a professor of physics and astronomy. He succeeded in preparing his paper on the Schwarzschild solution for publication at this time, while in the hospital, but not long after, he died of his disease at the age of 42.

Independently, and around the same time, Dutch physicist Johannes Droste worked out a similar exact solution to Einstein's field equations. Droste's mentor and professor, Hendrik Lorentz, saw to it that Droste's paper was published in a Dutch physics journal, and it appeared after Schwarzschild's paper, in 1917.[3] Perhaps because Einstein was a friend of Schwarzschild, he never gave credit to Droste for his independent result. The exact solution of Einstein's field equations is universally known as the *Schwarzschild solution*. In view of the historical situation, perhaps it should be called the *Schwarzschild-Droste solution*.

In 1916, Hans Reissner published a solution of Einstein's field equations combined with James Clerk Maxwell's field equations for electromagnetism, obtaining a static, spherically symmetric solution for a charged particle.[4] Independently, in 1918, Gunnar Nordstrøm published the same solution.[5] This static, spherically symmetric solution has now been identified as an electrically charged black hole. But in reality, astrophysical bodies, including black holes, are electrically neutral. There can be an excess charge in the body, but it will be negligible and does not affect the spacetime geometry.

THE EVENT HORIZON CONUNDRUM

The Schwarzschild radius of a body is given mathematically as two times Newton's gravitational constant, multiplied by the mass of the object, divided by the square of the speed of light ($2GM/c^2$). If a star shrinks to this radius, it triggers the formation of a so-called *event horizon* of the black hole—the one-way boundary where everything falls into the black hole and nothing escapes (Figure 1.2).

3. J. Droste, "The Field of a Single Centre in Einstein's Theory of Gravitation, and the Motion of a Particle in that Field," *Proceedings of the Royal Netherlands Academy of Arts and Science*, **19** (1), 197–215 (1917).

4. H. Reissner, "Über die Eigengravitation des elektrischen Feldes nach der Einsteinschen Theorie," *Ann. Phys. (Germany)*, **50**, 106–120 (1916).

5. G. Nordstrøm, "On the Energy of the Gravitational Field in Einstein's Theory," *Proc. Kon. Ned. Akad. Wet.*, **20**, 1238–1245 (1918).

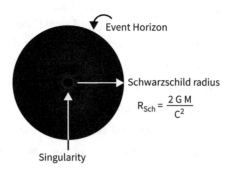

Figure 1.2. Schwarzschild black hole. Credit: Website ScienceBlogs.com (Cornell University)

At the birth of a black hole, a star that collapses under its own mass forms an event horizon. Whatever falls through or across the event horizon as the star collapses disappears permanently. This idea of a collapsed star forming an event horizon emerges in a concrete way from Einstein's gravity theory. In 1939, Robert Oppenheimer and his student Hartland Snyder published a solution showing how a collapsing star would form an event horizon, preventing light from escaping.[6] This was a major step beyond the discovery of the Schwarzschild and Droste solution.

One of two bizarre properties of the black hole event horizon is that, for astronauts falling through the event horizon, they will not experience any physical change. However, as a result of the choice of Schwarzschild coordinates, space becomes time, and the astronauts always move irreversibly forward in time, in contrast to being able to more forward or backward or in any other direction in three-dimensional space. Thus, an astronaut, having fallen through the event horizon, is drawn inexorably in a short time to the singularity at the center of the black hole, where matter is compressed to infinite density and time ends. This singularity is an inevitable mathematical consequence of the Schwarzschild solution. Here, there is no turning back! All the matter that falls through the event horizon, including astronauts, advances with time toward the singularity at

6. J.R. Oppenheimer and H. Snyder, "On Continued Gravitational Contraction," *Physical Review*, **56**, 455–459 (1939).

the center. Inside the final, quiescent state of the classical general relativity black hole, there is no matter at all between the central singularity and the event horizon. However, before the sorry astronauts reach the singularity, they will be "spaghetticized" by the strong gravitational tidal forces inside the black hole. Similar to the moon acting on Earth's oceans to produce tides, an astronaut will be subjected to a differential pull of gravity between the singular center of the black hole and the weaker gravitational force at the event horizon. The part of the astronaut's body that is leading the descent toward the singularity, whether head or feet, will be squashed and stretched like spaghetti.

However, this spaghetticizing happens only in *ordinary* black holes with masses similar to our Sun. A supermassive black hole, such as the one at the center of our galaxy, will have a density of matter equivalent to about a thousand times the density of water, so the tidal forces will be weak. There, adventurous astronauts can live reasonably comfortably until they are drawn into the singularity at the center of the supermassive black hole, which will finally annihilate them.

The second bizarre property of the black hole event horizon is that a distant observer will never see the event horizon form, for an infinite amount of time is required for light to escape. This is one of those extreme effects of general relativity that are so hard to wrap one's mind around. The person falling through the event horizon will do so in finite time, without even detecting the presence of the event horizon. This paradox is part of the nature of relativity theory.

Already, in Einstein's special theory of relativity of 1905, which did not yet include gravitation as a spacetime phenomenon, there occurs what is called the *twin paradox*. This has to do with the speed of one observer compared to another in the absence of gravity. A twin at rest on Earth finds that his clock is moving faster than the clock of his twin on a spaceship moving close to the speed of light. The twin who returns to Earth will be biologically younger than the one who stayed behind.

This weird situation comes about because light has the same speed for all observers. All observations of clocks and rulers depend on electromagnetism—in other words, light. One observer's measurement of time will depend upon their velocity relative to another observer. This is at the heart of relativity theory and has been verified experimentally by comparing the times on very accurate cesium clocks at rest on Earth and flying in airplanes.

We can now relate the twin paradox to the black hole event horizon paradox—that is, with the frame of reference of the distant observer and the reference frame of the astronaut falling through the black hole event horizon. The distant observer's clock measures what is called *coordinate time*, whereas the internal clock of the astronaut falling into the black hole measures what is called *proper time*. This does not imply a value judgment, but is simply terminology that has evolved over the years in relativity theory. The proper time of the astronaut falling toward the event horizon slows down and reaches zero at the event horizon, whereas the time measured by a distant observer who is moving away from the event horizon speeds up, and the time measured by a distant stationary observer remains the same; it does not slow down or speed up. The experimental verification of relativity is so accurate that these so-called *paradoxes* are not, in fact, paradoxes at all, but are natural physical consequences of a well-established theory.

GETTING COORDINATED

A problematic characteristic of a black hole is the mathematical singularity at the center. In the original form of the Schwarzschild solution, there also exists a mathematical singularity at the Schwarzschild radius corresponding to the event horizon. However, by choosing a different coordinate system, which covers all of the black hole spacetime, this second singularity at the event horizon can be removed.

What do we mean by a *coordinate system*? The normal definition of a Cartesian coordinate system is the set of three spatial coordinates x, y and z, and the one time coordinate t, used to label the position of an event in space and time. Other examples of coordinate systems are the Mercator cylindrical map projections of Earth and the longitude and latitude coordinates used to label positions on Earth. The singular event horizon at the black hole occurs because of a poor choice of coordinate system. When we describe Earth using longitude and latitude coordinates, a mathematical singularity occurs at the North and South Poles, where longitude becomes undefined. However, this is again simply the result of the mathematical coordinate system we have chosen to describe the geometry of Earth. These unphysical singularities at the poles are analogous to the mathematical singularity at the event horizon in the Schwarzschild coordinate system for black holes.

Early 20th-century physicists, including Einstein, misunderstood the significance of the singular event horizon as described by Schwarzschild and Droste. This was one reason that Einstein, and later Arthur Eddington, argued against the existence of black holes. Eddington, in fact, famously claimed that there must exist a law of nature that would prevent the formation of an event horizon and a black hole. Later, physicists realized that the singular behavior at the event horizon was simply a result of the choice of coordinates.

As it turned out, Eddington provided a solution to the problem of the singular event horizon in a short article in *Nature* in 1924. He compared Cambridge philosopher and mathematician Alfred North Whitehead's gravitational theory to Einstein's theory of gravitation (Figure 1.3), and to do this, he introduced a new coordinate system in which he transformed the metric of the Schwarzschild solution to a different but equivalent form in the new coordinate system. The term *metric* refers to the metric tensor in non-Euclidian geometry, which determines the infinitesimal distance between two events in spacetime.

Now the singularity that occurred at the event horizon for the Schwarzschild radius had been removed, although Eddington did not alert the reader to the fact that he had solved the singular event horizon problem. Much later, in 1958, David Finkelstein, at the Georgia Institute of Technology, used the "Eddington coordinates" to explain that the event horizon was a nonsingular, one-way boundary or membrane from which light could not escape. Afterward, the Eddington coordinates were renamed the *Eddington-Finkelstein coordinates*. At about this same

Figure 1.3. Sir Arthur Eddington and Albert Einstein. Credits: Eddington photo by Walter Stoneman, public domain from the Library of Congress Prints and Photographs Division, Washington, D.C; Einstein photo from Wikipedia Commons (Ferdinand Schmutzer, photographer, 1921)

time, a physicist at the University of California at Los Angeles, Christian Fronsdal, independently published a version of the Eddington-Finkelstein coordinates, and Roger Penrose, at Oxford University, subsequently converted the Eddington-Finkelstein coordinates into a form valid for light rays crossing the event horizon.

Decades later, building on the collapsing star solution derived by Oppenheimer and Snyder in 1939, Princeton mathematician Martin Kruskal and George Szekeres, a Hungarian-Australian mathematician, further resolved the issue of the singular black hole event horizon. In 1960, they independently showed that there exists a coordinate system description of a black hole that is smooth and regular—that is, without singularities—until you reach the essential singularity at the center of the coordinate system.[7] In other words, they also did away with the singularity

7. M.D. Kruskal, "Maximal Extension of Schwarzschild Metric," *Physical Review*, **119**, 1743–1745 (1910); G. Szekeres, "On the Singularities of a Riemannian Manifold," *Publ. Mat. Debrecen*, **7**, 285–301 (1960).

at the event horizon, just like Eddington had. This was an important discovery, because if the black hole event horizon had continued to be considered a singular surface with infinite density, then even the most enthusiastic black hole protagonists would have found this unacceptable, because such a surface is physically meaningless.

When Kruskal showed famous Princeton professor John Wheeler his mathematical solution for the black hole, Wheeler urged him to publish it as a paper, despite the fact that the mathematician Kruskal did not consider this work particularly significant. Wheeler insisted, and helped Kruskal write a paper on this solution. Although the Eddington-Finkelstein coordinates covered only part of the spacetime describing a black hole with a nonsingular event horizon, the Kruskal-Szekeres coordinates covered all of the black hole spacetime. However, although the black hole solution in the Eddington-Finkelstein coordinates described a static black hole, independent of time, the Kruskal-Szekeres solution depends on both time and space.

Another important development, in 1964, was Roger Penrose's discovery of what is now called the *Penrose diagram* (Figure 1.4), which explained in mathematical detail the nature of the Kruskal completion of spacetime. Penrose compressed infinite distances into a finite diagram by what is called a *conformal mapping of coordinates*, allowing for a much clearer understanding of the black hole. The Penrose diagram allows us to picture the essential features of a Schwarzschild black hole in spacetime.

DO BLACK HOLES REALLY EXIST?

In 1939, Einstein published an article in which he attempted to prove that a black hole could not form.[8] Rather than working with a single collapsing star, Einstein pictured a spherically symmetric and nonrotating

8. A. Einstein, "On a Stationary System with Spherical Symmetry Consisting of Many Gravitating Masses," *Ann. Math. (U.S.A.)*, **40**, 922–936 (1939).

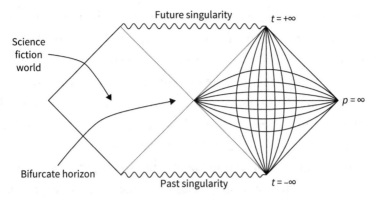

Figure 1.4. Penrose diagram of the Schwarzschild black hole. The two-dimensional Penrose diagram has a vertical time coordinate t and a horizontal spatial coordinate ρ *(rho)*, and ρ=infinity is spatial infinity. The black hole is the top center triangular region of the diagram, bounded by the future singularity above. Similarly, the white hole is the bottom center triangular region, bounded by the past singularity below. The left side of the diagram, denoted by "Science Fiction World," describes a hypothetical alternative universe. If the black hole turns into a wormhole, it is to this alternative universe that one might escape through the wormhole. No light can escape from a black hole, and all light can escape from the white hole. The spacelike essential singularities of the black and white holes are shown as wiggly lines and represent future and past singularities. During the gravitational collapse of a star, an external, distant observer will only observe horizon formation at future infinity, whereas an observer who falls through the horizon will be unaware of the existence of the horizon. Credit: Viktor T. Toth

conglomerate of stars, and showed that as the system contracted to the point when the individually rotating stars reached their critical escape velocity, they could no longer leave the system. He assumed there were internal forces and pressure that prevented gravity from allowing the stars to be below their critical escape velocity. At this point, he showed that if the stars were to collapse into a dark object, they would be moving at a speed greater than the speed of light, which is not permitted by his special relativity theory. He used this calculation to support his objections to the existence of dark objects, which are now called black holes.

Einstein made the mistake of supposing there were internal forces and pressures strong enough to prevent the collapse of the initial conglomerate of stars into a Schwarzschild black hole. Einstein went to his grave in 1955 not believing in the existence of black holes.

The problem with Einstein's attempt to prevent the formation of black holes was his not using quantum statistics, which he actually invented in 1925. These statistics followed from his investigation of the collective modes of molecules at very low temperatures, and his prediction of Bose-Einstein condensates, following the seminal work of Satyendra Nath Bose. The quantum pressure from the degenerate electron and neutron gases can, in most cases, prevent the collapse of a star to a black hole. Some months later, following Einstein's published paper, Oppenheimer and his students Hartland Snyder and George Volkoff calculated how a sufficiently massive star would collapse under its gravitational forces. In the seminal paper by Oppenheimer and Snyder, published soon after Einstein's in 1939, they showed for the first time how black holes formed. They did not refer to Einstein's paper claiming that black holes cannot form.

We now know, thanks to this work, as well as work by Lev Davidovich Landau, Subrahmanyan Chandrasekhar, John Wheeler, Masami Wakano, and Kent Harrison, that the pressures resulting from electron and neutron gases are not sufficient to beat out the attractive force of gravity. What actually happens is that, before the stars' orbital velocities reach the speed of light, a horizon forms at the center and grows until it encloses the entire collapsing star. All the enclosed matter advances inevitably to the singularity.

Other physicists continued to improve our understanding of the mathematical solution describing a black hole, and the physics community began to accept that black holes really do exist. From the 1960s onward, a considerable body of physics research has now been published supporting the idea that black holes do exist in nature.

An important contribution came in 1963, when New Zealand physicist Roy Kerr pushed the consensus on black holes a step further. He discovered a solution to Einstein's field equations that described a rotating black hole.[9] This solution reduces to the Schwarzschild solution when the

9. R.P. Kerr, "Gravitational Field of a Spinning Mass as an Example of Algebraically Special Metrics," *Physical Review Letters*, 11, 237–238 (1963).

angular momentum or spin of the black hole is zero. Astrophysical bodies such as stars rotate and possess an angular momentum; they spin on their axes just like the planets and the Sun do. Kerr showed that black holes also rotate on their axes. It is now accepted that black holes rotate like tops or planets, just like the collapsed stars from which they were derived.

2

Thermodynamics, Quantum Physics, and Black Holes

In the early 1970s, Jacob Bekenstein, who was at that time a graduate student in John Wheeler's well-known relativity group at Princeton, proposed that black holes must have entropy—which is the tendency of a system to become disordered, and a measure of the information hidden within a system; otherwise they would violate the second law of thermodynamics. This law states that entropy always increases with time. He said that if you throw a lump of matter into a black hole, as well as increasing the matter in the black hole, the lump of matter will increase the black hole's entropy. If the black hole did not possess entropy, then you could not accept that the second law of thermodynamics is valid for a black hole. A body that has entropy should also have temperature, and any body with temperature radiates energy. Bekenstein's work stimulated physicists for the first time to be seriously concerned about the temperature of black holes. Entropy, temperature, and energy determine the basic properties of classical thermodynamics.

Stephen Hawking, at Cambridge University, refuted Bekenstein's proposal that a black hole without entropy would violate the second law of thermodynamics, noting that no information can escape the black hole's event horizon because of the extreme gravitation present there. Because Bekenstein's proposal says that the black hole has a temperature, this would contradict the consensus idea that a black hole does not radiate energy because of the event horizon radiation barrier. An object that has a temperature. such as a black stove, emits electromagnetic radiation, and thereby the temperature of the body decreases over time. Hawking stated that because a black hole does not emit any kind of radiation, because of the existence of the event horizon, it cannot have a temperature, and therefore cannot possess entropy.

ENTER QUANTUM PHYSICS

Almost all physicists agreed with Hawking's conclusion regarding Bekenstein's claim. An exception was Yakov Borisovich Zel'dovich, a well-known Russian astrophysicist at Moscow University who had played a significant role in the Soviet Union's development of a hydrogen bomb. He conceived of a way for a black hole to emit radiation. The mechanism for this radiation emission is based on quantum physics—in particular, on relativistic quantum field theory in which the vacuum is not just nothing, but consists of ethereal pairs of positively and negatively charged particles being created and annihilated so that the average temperature of the vacuum is zero. These so-called *virtual charged particles* have a nonzero average energy for the positively charged particle and a nonzero average energy for the negatively charged particle, which on average cancel one another out, so that the vacuum appears to be empty of matter. This virtual creation and annihilation of particles takes place in a time period allowed by Heisenberg's Uncertainty Principle—that is, without violating the conservation of measurable energy.

The Uncertainty Principle states you cannot simultaneously measure with certainty the position of a particle and its momentum. Moreover, you cannot simultaneously measure a particle's energy and the time duration of its quantum mechanical movement. Therefore, you can violate the conservation of energy for a particle in quantum mechanics, given that the interval of time is small enough during this event. So for spontaneously created positively and negatively charged particles in a vacuum, energy conservation can be violated, provided they annihilate each other and cease to exist within a sufficiently short time. These events are called *the creation and annihilation of virtual particles.*

Zel'dovich proposed that a spinning black hole has an atmosphere of fluctuating vacuum pairs of virtual particles, and the tidal forces of gravity near the black hole event horizon could be big enough to separate the positive and negative energy particles sufficiently to produce a small number of *real* particles such as electrons, which would then turn into an electron radiation field being emitted by the black hole. This picture can be generalized to any particle pairs, including pairs of uncharged photons and uncharged gravitons, which for the photons would produce electromagnetic radiation, and for the gravitons would produce gravitational wave radiation. Zel'dovich's mechanism was tightly connected to the spin, or angular momentum, of the black hole. He claimed that the atmosphere of vacuum fluctuations close to the black hole event horizon would cease when the angular momentum of the black hole was reduced to zero. In other words, when the black hole ceased to spin, the emission of radiation would cease.

In 1973, Stephen Hawking visited Moscow and discussed the physics of black holes with Zel'dovich and his collaborator Alexi Starobinsky. They discussed, in particular, the identification of properties of the black hole and its relationship to thermodynamics. Eventually, Hawking would be swayed to agree with Bekenstein that a black hole should have a temperature and thermodynamic properties.[1]

Since the onset of his disease, amyotrophic lateral sclerosis (ALS), while an undergraduate at Cambridge University during the 1960s, Hawking had begun to take a deeper interest in physics. As it turned out, he had a slow form of ALS, so there was hope for him to live with this health problem for years. During research that took place mostly in his mind, because he was already suffering from severe limitations in using his hand to write or do calculations on paper, Hawking developed a deeper understanding of the meaning of the event horizon of a black hole. Relativists had come up with the idea that there are two kinds of event horizons. One

---

1. For a detailed account of the history of this event, see Kip Thorne's *Black Holes & Time Warps: Einstein's Outrageous Legacy.*

is called the *apparent horizon* and the other is called the *absolute horizon*. The apparent horizon is dependent on the observer's position and motion. That is, a given observer may have one form of apparent horizon, whereas another observer has a different one. On the other hand, the absolute horizon is the same for all observers, regardless of their motion. However, recall that an external observer cannot see an event horizon form because it occurs in the infinite future, whereas observers falling through the horizon do not know of its presence either, although they eventually suffer a spaghetti-like stretching by gravitational tidal forces as they approach the singularity.

Hawking pictured two black holes coalescing such that their event horizons became one in a final black hole. This led him to his *area law*— namely, that the area of the final black hole event horizon is greater than the sum of the areas of the individual coalescing black hole event horizons. In other words, the final entropy has to be larger than the sum of each component black hole entropy to fulfill the second law of thermodynamics. This led Hawking to identify the area of the horizon of the black hole with entropy, because of the second law of thermodynamics, which states that the entropy or randomness of an object always increases. For Hawking, the entropy was all located at the surface of the event horizon, not in the internal volume of the black hole. This means that, in fact, the event horizon must never shrink.

Now Hawking was in a position to agree with Bekenstein's interpretation of black holes: that they had to have a temperature, because you can't have entropy without temperature, although there was still the problem that if the black hole has a temperature, then it must emit radiation. Bekenstein, as John Wheeler's graduate student, had come up with the result that the black hole entropy had to have a specific form—namely, it was proportional to the area of the black hole event horizon divided by the Planck length squared. The Planck length is a quantum physics fundamental length formed from Planck's constant h, the gravitational constant G, and the speed of light c.[2] Bekenstein's use of the Planck constant united quantum mechanics with the thermodynamics of black holes.

---

2. The Planck length is given by the square root of $(Gh/2\pi c^3)$, which equals 1.6×10^{-35} meters.

HAWKING RADIATION

During intensive research at Cambridge, Hawking eventually came up with the idea that a black hole had to emit radiation, and he believed this would occur independently of its spin or angular momentum. He used Zel'dovich's idea that the radiation had to be produced purely by quantum physics, using the notion of a fluctuating vacuum. He reasoned that a virtual pair of photons would split into two partners. One partner would have negative energy and the other, positive energy; the negative-energy partner would fall through the horizon into the black hole whereas the other would remain outside (Figure 2.1). The conservation of energy would be maintained during this process as a result of Heisenberg's Uncertainty Principle, which allows for the nonconservation of energy to occur virtually in the uncertain time allowed by the Uncertainty Principle.

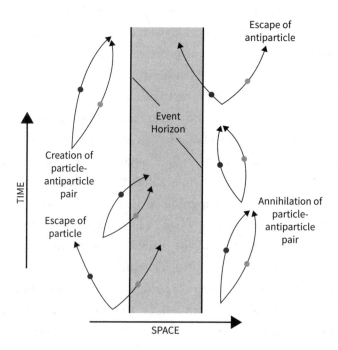

Figure 2.1. Hawking radiation. Credit: Stephen Dilorio, Union College, "A closer look at Hawking radiation."

The tidal forces at the event horizon would be strong enough to produce real photons from the virtual, fluctuating photons outside the horizon, and this would be seen by a distant observer as electromagnetic radiation. The negative-energy photon falling in would decrease the mass, area, and entropy of the black hole. Hawking generalized this idea to be true for any virtual pairs of particles, such as positrons and electrons, or neutrinos and antineutrinos, or gravitons. (The photon and graviton are their own antiparticle.) The resulting radiation was of a purely random nature, such as the radiation from a black body, which is just random noise. Hawking described these ideas in an article published in *Nature* in 1974.[3]

Meanwhile, back in Russia, physicists did not at first agree with Hawking's conclusions about black hole radiation. Both Zel'dovich and Starobinsky insisted that the emission of radiation by the black hole had to be tightly constrained by the spin, or angular momentum, of the black hole. However, after further research, they eventually came to agree with Hawking's result that a black hole emits radiation through vacuum fluctuations near the event horizon, regardless of whether the black hole is spinning.

Other physicists were still skeptical, however, and it took some years before there was general agreement that Hawking was correct. What came to be known as *Hawking radiation* was closely connected to an idea by Canadian physicist William Unruh, who had been a student of John Wheeler. Unruh described an astronaut in a spaceship that was kept at rest near the event horizon by a strong acceleration of the ship. This prevented the astronaut and ship from falling through the event horizon, and, Unruh claimed, it also triggered radiation emissions from near the event horizon. This idea stems from the fact that any accelerated charged particle produces radiation. On the other hand, observers freely falling through the event horizon and not subject to acceleration would not observe any radiation. Unruh's idea was related to Hawking radiation, but is a physically different phenomenon.

3. S.W. Hawking, "Black Hole Explosions?" *Nature*, **248**, 30–31 (1974).

ALICE AND BOB

The fact that for distant observers it takes an infinite amount of time to receive light from the black hole's event horizon implies that observers cannot watch unsuspecting astronauts fall into the black hole. Instead, from the vantage point of the distant observers, the astronauts would appear to be frozen just above the event horizon. Yet at the same time, the astronauts—independent of what the distant observers see, would be unaware of the existence of the event horizon as they fall through it. This bizarre conundrum of appearances to different observers has played a central role in attempts to understand black holes, and continues to puzzle theorists. It is a direct consequence of the relativity and observer dependence of time for strong gravitational fields. In fact, speaking of time, if the static observer, Bob, when watching Alice in a spacecraft approach the black hole event horizon, is able to observe her smartphone, he would notice that the time displayed on Alice's phone slows down as she approaches the event horizon and eventually comes to a stop as he watches her become frozen. This apparent stretching of time, too, is a consequence of the relativity built into Einstein's gravity theory.

The black hole saga has been greatly enriched by importing quantum mechanics into it. In particular, the two observers, Alice and Bob, like the particles and antiparticles in Hawking radiation, are quantum mechanically entangled. Thus, they illustrate a fundamental phenomenon in quantum physics: the reality of quantum entanglement of particles such as photons and electrons. Pairs of particles in quantum mechanics can share the same quantum state, which is known as a *superposition of states*, and they remain nonlocally "entangled" no matter how far apart they are. Measurements of particle spin have verified that for a pair of electrons distantly separated, when the spin "up" is measured for one electron, then it is instantaneously known that the spin of the other electron is "down." The second paired electron "knows" instantaneously, upon measurement, that its spin is down. This mysterious marriage of particles in quantum mechanics is nicknamed *Alice and Bob entanglement* (Figure 2.2).

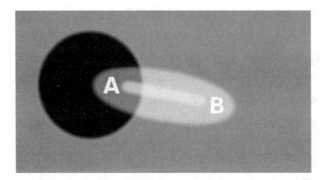

Figure 2.2. Quantum entanglement. There are two kinds of connections for Alice and Bob, with one inside and one outside a black hole: quantum entanglement and a wormhole, or a tunnel in spacetime. Credit: Martin Green.

This physical phenomenon was investigated in a famous paper by Einstein, Podolsky, and Rosen.[4] Einstein referred to this instantaneous pairing of the electrons as "spooky action at a distance," for it suggested it would require an infinite speed of light to communicate the information between the two electrons. In the modern interpretation of quantum mechanics, there is no violation of the constant, finite speed of light, because the phenomenon is a purely quantum mechanical one, unrelated to the classical speed of light. The quantum entanglement does not violate the finite, constant speed of light in special relativity. Moreover, causality is not violated because no information can be imparted from one entangled particle to another instantaneously. It is, rather, because the two particles were created as one unit.

In quantum mechanics, quantum entanglement comes about when Alice and Bob are created as one unit. So in a sense, they are always entangled, however far apart they may be in space. If Bob is outside the event horizon, then he can be entangled quantum mechanically with Alice inside the event horizon.

4. A. Einstein, B. Podolsky, and N. Rosen, "Can Quantum-Mechanical Description of Physical Reality be Considered Complete?" *Physical Review*, **47**, 777–780 (1935).

GRAPPLING WITH HAWKING RADIATION

We now arrive at the problem of how we could ever measure Hawking radiation. According to Hawking's calculations for the evaporation of a black hole, a black hole of 10 times the mass of the Sun would take about 10^{67} years to evaporate through Hawking radiation. This is more than about 10^{57} times longer than the age of the universe! Moreover, the temperature of the Hawking radiation is about 10^{-9} Kelvin, which is far too low to be measured. The temperature of the cosmic microwave background in the early universe is about 2.7 Kelvin.

On the other hand, it is possible that black holes were formed in the early universe when the density of matter and pressure were great and therefore gravity was very strong. These black holes would have a mass of about 10^{14} grams, similar to the mass of a large mountain. Such so-called *primordial black holes* could evaporate within the present age of the universe and, at the end, they would explode in a huge burst of gamma radiation. However, such primordial black holes have never been observed, after much effort to detect them. The bursts of gamma radiation associated with stellar objects in the sky can be explained by other astrophysical phenomena.

Why is it, then, that we believe in Hawking radiation? At this stage, the physics community believes purely on theoretical grounds that Hawking radiation exists. Our understanding of relativistic quantum field theory and vacuum fluctuations is consistent with Hawking's reasoning that black holes radiate. However, nothing has been empirically proved. Until we detect Hawking radiation directly, we must remain skeptical about its reality. The basic purpose of physics is to employ mathematics to predict physical phenomena that can be observed in nature. Without verification of the predictions through observations, they remain mathematical speculations.

In summary, the idea of Hawking radiation led Hawking to his famous discovery that a black hole is not actually strictly "black," but it evaporates by means of Hawking radiation. This resolved the entropy problem, as envisaged by Bekenstein, and we began to see the black hole as a normal physical body with a temperature and emitting radiation. The subsequent development of the thermodynamics of black holes led to the need for a quantum physics understanding of them, and—in particular—stimulated research on quantizing the gravitational field.

During the past three decades, the theoretical acceptance of black holes has increased to such an extent that few physicists question that black holes exist. The data coming in from the Event Horizon Telescope (EHT) and the Laser Interferometer Gravitational-Wave Observatory (LIGO)/ Virgo experiments are increasing our observational understanding of black holes.

THE INFORMATION LOSS PARADOX

Hawking raised a serious problem while developing his ideas of Hawking radiation. According to the fundamental equations of classical and quantum physics, any system developing in time can be reversed in a symmetric fashion. This is called *time reversal invariance*. In the case of information that enters a black hole, time reversal invariance is broken because any objects carrying information into the black hole must inexorably proceed in time to the singularity at the center of the black hole, from which there is no escape. This violates the time–symmetric invariance in quantum mechanics, and leads to a permanent loss of information inside the black hole. In a dramatic sense, information is lost in a black hole.

In our universe, we strongly expect that information of any kind, whether matter or electromagnetic radiation, cannot be destroyed or simply disappear. This is the basic concept of the conservation of matter, energy, and information. Yet, when anything falls into a black hole, it disappears and is lost. Recall that Hawking radiation is purely randomized thermal radiation, which is the opposite of ordered information, and has no information content, so Hawking radiation cannot resolve the problem of information loss. This is what is known as the *information loss paradox*. Physicists have devoted considerable effort to resolving this problem, with little success. Unfortunately, as with other aspects of the

Hawking radiation phenomenon, it is unlikely ever to be possible to confirm or deny experimentally whether information *is* indeed lost within a black hole. Because we believe in the consistency of quantum mechanics and relativity (including black holes), the suggestion that there is an inconsistency in these theories, as evident in the information loss paradox, continues to distress physicists.

ALTERNATIVES TO BLACK HOLES WITHIN GENERAL RELATIVITY: THE GRAVASTAR

Despite the consensus on the existence of black holes by the physics community, are there other solutions of Einstein's field equations in general relativity that could lead to an alternative to black holes? Could solutions be found that remove the event horizon, leaving an almost black star rather than a black hole in spacetime, and also remove the information loss paradox?

An observer hovering with the aid of a rocket outside the black hole event horizon and looking upward sees the frequency of light falling into the black hole—such as light from a distant star—to be shifted to the blue end of the electromagnetic spectrum. The same hovering observer, when looking down sees redshifted light from matter falling into the black hole. We note that in general relativity, light emitted from a massive body undergoes gravitational redshifting. This means that when a star is collapsing, a distant observer sees that the light emitted by the collapsing star is redshifted until the star forms an event horizon, when light can no longer escape.

Meanwhile, starlight falling into the black hole is blueshifted at the Schwarzschild radius, or the event horizon, of the black hole. In quantum mechanics, the energy of a photon is equal to Planck's constant times the frequency of the photon wave. Because the frequency of the photon hitting the event horizon of a black hole is infinitely blueshifted, the energy of the photon at the Schwarzschild radius is also infinite. This would indicate that the quantum physics at the black hole event horizon can lead to some serious physical consequences for the event horizon, such as the concept of infinite energy.

This problem led Pawel Mazur and Emil Mottola to seek a solution in quantum physics. In 2001, they introduced the idea of a *gravastar*, which is short for *gravitational vacuum star*.[5] This object has a quantum mechanical phase transition of matter at its Schwarzschild radius and in its interior. These are not black holes. Rather, they are stars that have an anisotropic, or uneven, surface and they have vacuum energy inside, rather than matter. The baryons that comprised the original progenitor star that underwent gravitational collapse now form a thin shell of baryonic matter around the vacuum energy interior of the gravastar. These dark objects do not have an event horizon, and there is no singularity at their center, as is the case with classical black hole solutions in general relativity.

In the gravastar, the quantum physics of the baryonic matter at the shell or crust surrounding the interior vacuum prevents the formation of a general relativity event horizon. A change of coordinates will not affect this quantum physics erasure of the event horizon.

Another quantum physics phase transition makes the interior of the gravastar into a Bose-Einstein condensate fluid, which avoids the central singularity of a general relativity black hole. In atomic physics, at low temperatures, it is possible to make particles form condensates that are fluid in form. In 1924 to 1925, Satyendra Nath Bose and Einstein predicted the existence of such condensate fluids. The Bose-Einstein condensate is a state of matter that is cooled to a temperature close to absolute zero (very near 0 Kelvin or –273.15 degrees Celsius). Such a cool, dilute gas is called a *superfluid*, and it has an extremely low density—namely, about 100,000th the density of normal air.

The gravastar vacuum energy is the quantum ground state energy of the Bose-Einstein condensate fluid. Recall that energy, in relativity, is equal to the mass times the square of the velocity of light. So the original matter making up the progenitor collapsing star is conserved upon gravitational

5. P.O. Mazur and E. Mottola, "Gravitational Vacuum Condensate Stars," *Proceedings of the National Academy of Science (US)*, **101**, 9545–9550 (2004).

collapse by the final gravastar. This final gravastar has a very high redshift of light, which makes it look "black" to an observer outside the gravastar. It looks just like a black hole in general relativity. It is approximately the same size as a black hole that has collapsed from a massive progenitor star, and the gravitational forces at the Schwarzschild radius would be the same as in a standard black hole, with its accompanying mass. I stress that the gravastar is a quantum physics and general relativity solution of Einstein's field equations, whereas a black hole is a purely classical solution of the equations, without taking into account any quantum mechanical physics.

DARK ENERGY STAR

In 2003, George Chapline, Robert Laughlin, and two other collaborators published an article in which they questioned the very existence of black holes in classical general relativity.[6] Similar to the ideas of the gravastar promoted by Mazur and Mottola, they invoked a quantum phase transition at the location of an event horizon, thereby evading the event horizon. They theorized that a phase transition occurring at the event horizon combined with the existence of a superfluid at the horizon would remove the event horizon in classical general relativity. He also theorized that the interior of the dark energy star consisted of a large vacuum energy associated with Einstein's cosmological constant.[7] In a speculative leap, this idea linked the interior of the dark energy star to the "dark energy"

6. G. Chapline, E. Hohlfeld, R.B. Laughlin, D.I. Santiago, "Quantum Phase Transitions and the Failure of Classical General Relativity," *International Journal of Modern Physics A*, **18**, 3587–3590 (2003).

7. In Einstein's first paper on cosmology published in 1917, he introduced the cosmological constant as part of his field equations. This constant produced a form of gravitational repulsion that balanced the gravitational attraction, producing a static, nonexpanding universe. Russian mathematician Alexander Friedmann subsequently published a paper in which he solved Einstein's field equations for cosmology, and demonstrated that the solutions predicted an expanding universe. Astronomer Edwin Hubble then established through observations that galaxies are receding from one another at a constant speed, and the concept of the expanding universe was born. Einstein, on the failure of his static universe, abandoned his cosmological constant and is supposed to have said that the cosmological constant was the "biggest blunder" of his life.

recently deduced by astronomers to be accelerating the expansion of the universe. Saul Perlmutter, Adam Riess, and Brian Schmidt, observing through supernovae data that the expansion of the universe was accelerating, suggested that "dark energy" and the cosmological constant were responsible.

In a classical general relativity black hole solution, all the matter of the initial star collapses to the singularity at the center of the black hole, and there is no matter left inside the black hole between the singularity and the surrounding event horizon. Quantum physics is brought in to change this picture radically. Astronauts falling into a gravastar would not experience the infinite spaghetticizing resulting from the infinite tidal forces produced by the black hole singularity because there *is* no singularity. Provided the astronauts can penetrate the thin shell of baryonic matter of the gravastar, then they will merely float around in the inner gravastar vacuum.

Several authors have investigated the physical properties of a gravastar, such as its required stability and whether it is possible to distinguish it experimentally from a standard black hole. Before the LIGO detection of gravitational waves, the conclusion was that there was no easy way to distinguish the two objects. However, the data for the first gravitational wave detection, GW150914, changed this situation. Other authors have published papers since 2016, including Niayesh Afshordi and collaborators at the Perimeter Institute and the University of Waterloo,[8] and Bob Holdom and collaborators at the University of Toronto,[9] showing that it is possible to use the LIGO data in the final ring-down phase of the merging black holes to discover whether the dense compact objects have a horizon. They

A. Einstein, "Kosmologische Betrachtungen zur allgemeinen Relativitätstheorie," *Preussische Akademie der Wissenschaften*, pt. 1, 142–152 (1917).

A. Friedmann, "Über die Krümmung des Raumes," *Zeitschrift für Physik*, **10**, 377–386 (1922).

8. J. Abedi and N. Afshordi, "Echoes from the Abyss: Tentative Evidence for Planck-Scale Structure of Black Hole Horizons," *Physical Review D*, **96**, 082004 (2017).

9. R. Conklin, B. Holdom, and J. Ren, "Gravitational Wave Echoes through New Windows," *Physical Review D*, **98**, 044021 (2018).

propose that a surface exists at a tiny distance, about the Planck length (10^{-33} centimeters) outside the horizon, that reflects electromagnetic and gravitational waves. Without such a surface, the black hole horizon would completely absorb all incoming electromagnetic and gravitational waves.

A surface very close to the horizon will affect the physics of the vibrational modes of the ringing compact object before it finally settles down to a nonperturbed object. By studying the LIGO data for the ring-down phase, physicists can determine whether so-called *echoes* appear in the data. The vibrational modes in the ring-down phase can be described by wave equations. Certain resonant modes called *quasi-normal modes* take on frequency-dependent distinct values, depending on the model used. In the case of a reflecting surface outside the event horizon, quasi-normal modes show periodic pulses of echoes corresponding to a kind of memory. A pure black hole without a reflecting surface at the event horizon would not reveal such echoes in the data. This suggests a new way of determining whether dense compact objects have event horizons, as is predicted in general relativity. To make definitive statements about this speculative idea, many more gravitational wave events are needed to provide sufficient critical data. Analysis of the ring-down phase data by the LIGO collaboration for the merging black hole events reveals no evidence to date for echoes in the ring-down phase.

The fact that the gravastar solution of Einstein's field equations combined with quantum mechanics cannot be easily distinguished from a black hole with current experimental data, including the advanced LIGO data and data from the ring-down phase, is a cautionary tale: the black hole story is still evolving. Clearly, the outcome of this tale is fundamental for the whole concept of black holes.

Despite the theoretical developments of the gravastar and the dark energy star, the vexing problem remains of whether the information loss paradox first proposed by Hawking in the early 1970s should motivate a serious revision of physics. Such a revision would involve either a revision of black hole physics or quantum mechanics. Because quantum mechanics is so well founded experimentally, it seems unlikely that the source of the problem lies within quantum mechanics. Similarly, will other paradoxical physical properties of a black hole, such as the existence of the event horizon, survive into the future?

There also exist classical solutions of alternative gravitation theories describing dark stars without event horizons. These dark stars would also resolve the information loss paradox. I discuss these theories in Chapter 9.

3

Stars and Black Holes

Since 2015, we have acquired stronger evidence that black holes exist because of the detection of gravitational waves produced during the coalescence of two black holes far away in space. Today's confidence is in stark contrast to the early history of black holes. Recall that Einstein, Eddington, and other luminaries did not believe in the existence of black holes. Einstein considered them an anathema, and to the end of his life he refused to believe they could exist.

Why was there a change in attitudes toward black holes or dark objects in space? One of the main reasons was the rapidly developing understanding of the constitution of stars, and how they are born and evolve.

THE LIFE STORY OF A STAR

Arthur Eddington, a staunch skeptic of black holes, was one of the world's most famous astronomers during the 1920s and 1930s. He was one of the few mathematical physicists who understood Einstein's theory of gravitation. In 1923, he published a masterful exposition of Einstein's theory in his famous book, *The Mathematical Theory of Relativity*. Eddington's support of Einstein's theory was radical at that time, because Newtonian gravitation, which had been extremely successful since 1687, was still the dominant theory. Yet the radical side of Eddington's personality as a scientist did not extend to believing in the ultimate fate of the collapse of a star.

Eddington was also famous for his book *The Internal Constitution of the Stars*, published in 1926. In this book he told a detailed story about how stars evolved from the clumping of cosmic gas clouds, and how they were kept stable under the influence of gravitation. This stability was the result of a balance between a star's internal gas pressure and the attractive force of gravitation. The gas in stars is made mostly of hydrogen atoms. The motion of these atoms in the gas produces heat, and the heat in turn produces pressure. This pressure counteracts the attraction of the gravitational pull of the atoms of the gas, and when a balance is reached, the star becomes stable.

An important question is: how are enough heat and pressure produced to make a star like the Sun shine for billions of years? We now know that the main source of stellar heat and pressure is thermonuclear fusion, which is also responsible for the energy in a hydrogen bomb. Eddington was one of the first to support the idea of nuclear reactions in stars being the source of their heat and pressure. However, developing the idea of this nuclear process mechanism took decades, and it was not fully understood until Hans Bethe's research on the nuclear physics of stellar thermonuclear fusion, which led to Bethe winning the Nobel Prize in 1967.

A star radiates away energy and continues to do so until it no longer has heat and the associated pressure to keep it stable. At this point, depending upon the mass of the star and the strength of its internal gravitational forces, the dying star can either explode as a supernova, with the core imploding and producing a neutron star or a black hole, or it can collapse to a cold and dead star like a white dwarf. In other words, as such an individual dying star cools down, it contracts as a result of gravity, and, unless a new form of pressure is produced to keep it stable, it will collapse under its own weight into a compact dense object such as a white dwarf, a neutron star, or a black hole.

It was not until the development of quantum mechanics in the 1920s by Bohr, Heisenberg, Pauli, Born, Schrödinger, and others, that an understanding was reached of what happens to this cold matter at the end of a star's life.

QUANTUM MECHANICS AND STARS

One of the key insights of the new quantum mechanics was that a particle like the photon could act like a particle and a wave at the same time. Einstein first showed this phenomenon with his photoelectric effect, and Prince Louis de Broglie developed the idea of the particle–wave duality. The particle–wave duality means the energy of a particle like a photon or an electron is inversely proportional to the wavelength of the particle. Here, the energy includes Einstein's relativistic mass–energy $E = mc^2$. The smaller the wavelength, the more energy the particle has. This means that for X-rays with very small wavelengths, you can have very large energies, sufficient even to penetrate flesh and bone, as Wilhelm Röntgen discovered in 1895.

In 1926, Ralph Fowler at Cambridge University investigated what happens when matter is compressed to the point where the motion of electrons is confined to tiny spatial units and, because of the electrons' mass–energy, the wavelengths cannot be larger than those units. Quantum physics had the electrons moving rapidly around inside these tiny cells and producing a new kind of pressure, a quantum pressure, in addition to the macroscopic heat pressure of dense matter. When the wavelength of the electrons became small enough, the resulting quantum mechanical energy and pressure far exceeded the ordinary heat pressure, and this new pressure could stabilize a star under its gravitational pull and keep it from collapsing. Fowler was a supervising professor at Cambridge of several physicists and astrophysicists who later became famous, including Paul Dirac.

In 1930, 19-year-old Subrahmanyan Chandrasekhar (Figure 3.1) had finished his undergraduate studies in Madras, India, and began studying the new quantum mechanics. He had been awarded a scholarship to attend Trinity College Cambridge for his PhD studies. Arthur Eddington was a Fellow of Trinity College, and Chandrasekhar had studied his book *The Internal Constitution of the Stars*, so he was anxious to meet the great astronomer. During the 18 days it took for his steamship to reach Southampton, he devoted his time to attempting to understand what

Figure 3.1. Subrahmanyan Chandrasekhar. Credit: Al Jazeera, from Getty Images

would happen to highly compressed matter such as a white dwarf after it radiated away all its energy. The result of his calculations was that a white dwarf could not exceed a maximum mass of 1.4 times the mass of the Sun, implying that if the mass were larger and the gravity strong enough, then the star would implode into either what we now call a neutron star or a black hole. He drafted an article describing his results. His conclusion was that, based on the work of Fowler, the cold star could be stabilized by this new quantum electron degenerate gas pressure. Here, the word *degenerate* refers to the fact that it is the lowest energy state of matter, in which individual atoms lose their identities. This mass limit of 1.4 solar masses came to be known as the *Chandrasekhar mass limit*.

At Trinity College, Chandrasekhar met Eddington and discussed his new ideas, and he also contacted Fowler and gave him his article on the quantum stability of a star. Fowler was not able to understand the derivation of Chandrasekhar's maximum mass for a white dwarf, so Chandrasekhar submitted the article himself to *The Astrophysical Journal*, and it was published in 1931.[1] Eddington recognized the importance of

1. S. Chandrasekhar, "The Maximum Mass of Ideal White Dwarfs," *Astrophysical Journal*, **74,**

Chandrasekhar's ideas about the fate of a cold star and took an interest in his work, and encouraged him to pursue this line of research further.

WHITE DWARFS, NEUTRON STARS, AND BLACK HOLES

The physics community did not respond with any interest to Chandrasekhar's findings on the white dwarf's maximum mass, so he turned to other research and to completing his PhD at Cambridge. Three years later, having completed his PhD, he returned to more comprehensive research of his white dwarf maximum mass result. Previously, he had only derived his result for low-density white dwarfs, whereas equations of state—namely, the relationship between pressure and density—could refer to much denser white dwarfs and required computer calculations. The equation of state for white dwarf matter had been derived in 1930 by English physicist Edmund Stoner at Leeds University, and independently by Wilhelm Anderson at Tartu University in Estonia. This equation of state could describe a white dwarf electron degeneracy gas for much denser white dwarfs. To solve the problem of the resistance of the degenerate electron gas to the gravitational force inside the white dwarf, Chandrasekhar had to solve a differential equation describing the stability of a white dwarf against the gravitational pull inside it. Fortunately, Eddington possessed a desktop mechanical calculator called a *Brunsviga calculator*, which was an elementary computer during the 1930s. This primitive computer allowed Chandrasekhar to perform long and tedious calculations that finally solved the problem of white dwarf stability.

In general, a white dwarf is a very dense star, with a volume comparable to that of Earth and a mass comparable to the Sun. Eddington himself was an expert on these astrophysical objects. Chandrasekhar studied the white dwarf called *Sirius B*, a faint companion star to the brightest star in the sky, Sirius, in the constellation Canis Major. Sirius and Sirius B are the sixth and seventh nearest stars to Earth, being about 8.6 light years away. Through telescopic observations, the mysterious white dwarfs had been

81–82 (1931).

discovered to have a very high density of matter—far greater than humans could ever encounter on Earth. According to the most recent observations and calculations, Sirius B has a mass of about 0.978 times the mass of the Sun and a circumference of only 36,700 kilometers, compared to the Sun's circumference of 4,366,813 kilometers. The modern value of the density of Sirius B is 4 million grams per cubic centimeter, compared to Earth's density of 5.5 grams per cubic centimeter and the Sun's density of 1.4 grams per cubic centimeter. The density of matter on Sirius B is difficult to imagine: one teaspoonful would weigh 16 tons, a dense white dwarf indeed!

What happens when a star of such high density collapses under its own weight? Both Chandrasekhar and Eddington were forced to consider this problem. According to Eddington, the high density of the white dwarf Sirius B created a serious conundrum. He attempted to explain the stability of Sirius B by the standard argument that the heat produced by the atoms making up the matter of Sirius B could cause enough pressure to keep the star stable. As the star radiates away energy, it shrinks, and then we are faced with the question of what happens at the end of the star's evolution. By Eddington's calculations, it could not eventually form a stable object and it would have to collapse—a fate he would not accept.

Chandrasekhar solved the problem of white dwarf stability by invoking his quantum mechanical degenerate gas pressure. This new pressure could take over after all the heat of the star had been dissipated by radiation. According to his calculations, this new pressure could counteract the attractive force of gravity. Thus, Eddington's heat mechanisms for stability had failed to explain the evolution of white dwarfs. Chandrasekhar's electron degenerate gas mechanism was based on the new quantum mechanics and was unknown to Eddington at the time he considered the problem of white dwarf stability. Chandrasekhar knew about this quantum mechanical gas through Professor Fowler.

This now raised the question: what would happen if you had a star with a mass greater than the so-called Chandrasekhar mass limit? Chandrasekhar was invited to present a paper about his research on the quantum mechanical mass limits of white dwarfs at a Royal Astronomical Society meeting at Burlington House, London, in 1935. Prior to the meeting,

Chandrasekhar learned that Eddington intended to give a talk following his talk. In Chandrasekhar's talk, he explained how he had derived the mass limit for white dwarfs using special relativity, Newtonian gravity, and the quantum physics pressure derived from the degenerate electron gas of the white dwarf. Chandrasekhar's concluding implication was that if a star had a mass of more than 1.4 solar masses—that is, it was more massive than a white dwarf—it would collapse to a dense dark object. In his talk that followed, Eddington discussed Chandrasekhar's results, but then argued that his treatment of special relativity and gravity could not be correct because it would lead to the absurd result that a star would collapse to some unknown object—namely, what we now call a black hole—and this was unacceptable. Eddington asserted that Chandrasekhar's conclusions had to be wrong.

The astronomers at the meeting went along with Eddington's arguments because Eddington was one of the most famous astronomers of his time and, besides, how could Chandrasekhar's predicted fate of a collapsing star be correct? It seemed too absurd to accept. Eddington never gave up his position on this issue, although he did respect Chandrasekhar's ability as an astrophysicist. Chandrasekhar's feelings were hurt and he decided not to pursue this line of research any longer because it would be detrimental to his career. He switched his research to other more conventional problems in astrophysics. Ironically, this early research was a main part of the reason the Nobel Committee awarded Chandrasekhar the physics prize ("for his theoretical studies of the physical processes of importance to the structure and evolution of the stars") in 1983 together with American experimentalist William Fowler (no relation to Chandrasekhar's Cambridge mentor, Ralph Fowler).

ENTER NUCLEAR PHYSICS

Is there any way to prevent a massive star from collapsing to a black hole? Quantum physics produced for us the degenerate electron gas and its pressure, which prevents a medium-size star like the Sun from collapsing to a black hole but instead produces a white dwarf, which is a cold, dead star. Now we are faced with the possibility that a heavy-enough star could go beyond the graveyard of a white dwarf and produce a black hole. If this were true for any initial star with a big enough mass, then Einstein and Eddington and their colleagues would be wrong, for nature would permit what appeared to them to be a ridiculous physical phenomenon. Answering the question of whether the formation of a black hole was inevitable or preventable would require the energies and focus of creative physicists as well as the evolving nuclear physics of the early 1930s.

At that time, the atom was believed to be a cloud of negatively charged electrons buzzing around a positively charged nucleus. In the case of hydrogen, it consisted of one proton and one electron, the most elementary atom. At this time, during the 1930s, Rutherford was trying to understand the nature of the atomic nucleus. To make the nucleus conform to quantum physics, he proposed the "neutron," an electrically neutral particle with about the same mass as the proton. To make the nucleus stable, he theorized there was a new force in nature called the *nuclear force* that held the protons and neutrons together. Furthermore, the neutron was responsible for the new quantum physics degeneracy pressure, and it made the whole nucleus a stable object. In February 1932, James Chadwick, who was a member of Rutherford's experimental team, discovered the neutron. He succeeded in bombarding an atomic nucleus, releasing a large number of neutrons.

This event caught the eye of Fritz Zwicky in Pasadena, California. He was an astronomer, born in Switzerland, who had been brought to the California Institute of Technology (Caltech) by Robert Millikan, who was head of the physics department. Zwicky was a somewhat belligerent person who claimed he possessed a creative genius; he was not popular with his colleagues. In the Caltech astronomy department was Walther Baade, a German who was a member of the Mt. Wilson Observatory and a brilliant observational astronomer. Both Baade and Zwicky were native German speakers, and they began a serious collaboration. A species of stars called *novae* flared up in our Milky Way galaxy and, accounting for their distance from Earth, produced a light much brighter in absolute terms than the light from the Sun. Zwicky became fascinated with these objects and soon latched on to another phenomenon—namely, a much more significant flaring of light in galaxies that he correctly perceived were distant from our own Milky Way. Baade had been studying these objects through the Mt. Wilson telescope with its 2.5-meter reflector, and Zwicky became an avid thinker about these objects. Baade and Zwicky coined the term *supernovae* for these dramatic objects.

The question arose at the time: what could produce such enormous amounts of energy? Zwicky, even with his limited knowledge of nuclear physics and theoretical physics, correctly hypothesized that the supernovae were produced by exploding stars. The explosion was the aftereffect of an implosion that produced what he theorized was a neutron core—an inner core composed almost entirely of neutrons. When the star imploded, its protons and electrons combined to form neutrons. Zwicky was able to make a rough estimate of the amount of energy that this implosion to a neutron core would produce: about 10^{42} joules, in a matter of seconds. (A one-megaton atomic bomb produces about 10^{15} joules.) Baade and Zwicky theorized that the release of this energy would cause a large amount of X-rays and ultraviolet light radiation to be emitted in an explosive manner.

But since the 1960s, we have known this was not correct. The radiation from the explosion of a star consists primarily of neutrinos, because a proton and an electron are converted into a neutron and a neutrino, leaving neutrons while the neutrinos escape. However, important features about the supernovae were correctly deduced by Baade and Zwicky. We also know today that supernovae are 10 billion times more luminous than our Sun.

NEUTRON STARS

We now come to the properties of the neutron star core of the supernova. Zwicky surmised that the supernova implosion would shrink the circumference of the neutron core to about 100 kilometers, compared to the many thousands of kilometers of the original star. It would also produce a mass for the core that would be about equivalent to the mass of the Sun, making it a superdense object. Baade and Zwicky gave a talk at Stanford University in December 1933 and published an abstract of it in *Physical Review*.[2] During the talk, they theorized that cosmic rays, which had been extensively studied by Robert Millikan, were mainly produced by these supernovae explosions occurring in outer space far from our Milky Way galaxy. They also said supernovae represented the transition of ordinary stars into *neutron stars*, stellar objects of extremely closely packed neutrons.

Although the physics community welcomed the explanation of supernovae provided by Baade and Zwicky, they were not happy with Zwicky's theorizing about neutron cores, which were now called *neutron stars*. One of the reasons is that Zwicky's explanations were limited in details. The physics community had merely to trust the power of his physical intuition. As it turns out, Zwicky's ideas about neutron stars were incredibly prescient. The idea of an ordinary star turning into a neutron star was confirmed observationally during the late 1960s with the discovery of a pulsar, a spinning, magnetized neutron star at the core of an exploded supernova. (All pulsars are neutron stars. However, not all neutron stars are pulsars.)

During the 1930s, it was still not understood how the Sun could provide enough heat and light to maintain itself and the solar system for more than 5 billion years without radiating to a cold star. Recall that Eddington had proposed that the heat in the Sun was produced by nuclear burning or nuclear fusion. However, nuclear physics was not advanced enough during the late 1920s and early 1930s to justify making this, as it turned out, correct statement.

In 1931, Russian theoretical physicist Lev Landau at Moscow University came up with the idea that he could use Zwicky's proposal of a neutron core to explain the Sun's longevity. He theorized that a neutron core at the center of the Sun would attract atoms, and when they fell into the neutron

2. W. Baade and F. Zwicky, "Supernovae and Cosmic Rays," *Physical Review*, **45**, 138 (1934).

Figure 3.2. J. Robert Oppenheimer. Credit: Wikipedia, from a Los Alamos publication

core, they would produce enough kinetic energy to heat the Sun. However, because the neutron had not yet been discovered, the idea was considered speculative. Landau took up the idea again in 1937 and 1938, after the neutron had been discovered (in 1932, by Chadwick). According to Landau's calculations, the neutron core, together with Newtonian gravity, could produce enough heat to sustain the Sun. To prevent the neutron core from deforming the Sun observationally, the mass of the neutron core had to be sufficiently light.

At Berkeley, Robert Oppenheimer (Figure 3.2) together with his student Robert Serber pondered this publication by Landau, and decided to investigate whether the idea of a solar neutron core could be physically viable. They discovered that, theoretically, a neutron core in the Sun could provide enough heat to prevent the Sun from radiating away too quickly. On the other hand, they found that the lighter mass of the neutron core that Landau required to make his idea work was not physically realizable. They published a critique of Landau's neutron core in the *Physical Review*.[3]

---

3. J.R. Oppenheimer and R. Serber, "On the Stability of Stellar Neutron Cores," *Physical Review*, **54**, 540 (1938).

Oppenheimer was now intrigued with the whole notion of the neutron core or neutron star and began investigating how massive such a neutron star could be before it collapsed under its incredible density to a dark object. If stars of any mass compatible with stellar evolution could be prevented by their internal gravitational attraction from collapsing to a black hole, then Einstein, Eddington, and others would be vindicated in their protest that Nature could not allow such a physical phenomenon as a black hole.

Oppenheimer collaborated with his student George Volkoff from Toronto, who had emigrated from Russia in 1924. Using calculations published by Richard Chase Tolman in his book *Relativity, Thermodynamics and Cosmology*, Oppenheimer and Volkoff found a solution for a neutron star in which they replaced the equation of state—the relation between pressure and density—for electrons appropriate for white dwarfs with an equation of state for neutrons.[4] Nuclear physics explained how the neutron degenerate gas pressure would counteract the attractive force of gravity, allowing the star to become a very dense neutron star, and preventing it from collapsing to a black hole.

From rough calculations, Oppenheimer found that a neutron star with a mass greater than six times the mass of the Sun could not support itself against the gravitational attraction and would have to collapse to a black hole. Oppenheimer and Volkoff refined the calculations that give the maximum mass of a neutron star and discovered the maximum mass would not, after all, be six times the mass of the Sun, but around three times the mass of the Sun. They were forced to use general relativity in their calculations because the circumference of a neutron star was some hundreds of kilometers, compared to the circumference of a white dwarf of about 10,000 kilometers, so the force of gravity was too strong to be described by Newtonian gravity. This was a significant result. It was the first strong indication that combining nuclear physics and general

4. J.R. Oppenheimer and G. Volkoff, "On Massive Neutron Cores," *Physical Review*, 55, 374–381 (1939).

relativity could not prevent massive-enough stars from collapsing to a black hole. The dreaded black hole, after all, seemed inevitable.

Meanwhile, Oppenheimer's criticism of Landau's neutron core mechanism—that it would not work to keep the Sun hot—left open the question of what *did* keep the Sun hot? This question was taken up by Hans Bethe and Charles Critchfield, who applied the developing nuclear physics to the problem.[5] Their calculations showed that Eddington's idea that the heat of the Sun was produced by nuclear fusion was correct. Later, John Wheeler in collaboration with B. Kent Harrison and Masami Wakano investigated how to explain the equation of state for neutron stars using up-to-date nuclear physics.[6] Their calculations confirmed that a neutron star with a mass greater than about three times the mass of the Sun had to collapse to a black hole, as discovered earlier by Oppenheimer and Volkoff. The biggest mass of a stable neutron star that has actually been *observed* is twice the mass of the Sun, which is compatible with the Wheeler-Harrison-Wakano equation of state and relativistic calculations based on the Oppenheimer-Volkoff equation. Of course, more massive neutron stars may be observed in the future because they are theoretically possible, according to general relativity.

From all of this, we can conclude that if there is no other physical mechanism besides neutron star degeneracy pressure and nuclear physics to prevent a neutron star with a mass bigger than three solar masses from collapsing, then we must confront the fact that black holes really exist in nature. The reader should note at this point that all of these results are based on theoretical inferences, assuming that Einstein's general relativity and nuclear physics are correct descriptions of nature. It would be decades before there would be stronger observational evidence for the existence of black holes.

5. H.A. Bethe and C.L. Critchfield, "The Formation of Deuterons by Proton Combination," *Physical Review*, **54**, 248 (1938).

6. B.K. Harrison, M. Wakano, and J.A. Wheeler, "Matter-Energy at High Density: End Point of Thermonuclear Evolution," *La Structure et l'Evolution de l'Univers*, Onzième Conseil de Physique Solvay, 124 (1958).

We now have knowledge about the possible maximal mass of a stable star. However, the formation of a black hole will occur through gravitational collapse. What was the status of knowledge of gravitational collapse during the 1960s? This was a decade in which there was a surge of fruitful new research being performed to understand the nature of the gravitational collapse of stars.

COLLAPSING STARS AND BLACK HOLES

Oppenheimer and Snyder had derived an idealized solution of stellar gravitational collapse in 1939. They assumed the star had uniform density with no pressure and that it was spherically symmetric. Oppenheimer, in a rough calculation before he and Snyder performed the remarkable mathematical feat that solved the problem, intuitively guessed that the issue of spherical symmetry was not important, even though the star would be rotating and would produce some distortion of spherical symmetry. Moreover, when the degenerate quantum gas pressure and heat pressure had finally been overcome by the sheer weight of the star, the gravitational collapse would take place dynamically, as if there was no pressure at all against gravity. The main discovery of the Oppenheimer-Snyder paper published in *Physical Review* was that a star would collapse through its critical Schwarzschild circumference, forming what we now call an event horizon—a contracted sphere at the edge of the black hole where the gravitational pull is strong enough to prevent even light from escaping.

A static observer sitting outside, watching the collapse of the star, would see light emitted from the collapsing star shift toward the red end of the spectrum, because it is shrinking rapidly away from us. When the star reaches the event horizon during collapse, the redshift becomes infinite and no light escapes from the black hole.

The relativity community during the '60s was skeptical about the assumptions in Oppenheimer and Snyder's calculations. In particular, much had been learned about the nuclear physics taking place internally in stars. The research that went into creating the uranium and plutonium bombs during World War II increased knowledge about nuclear reactions considerably. Moreover, the production of the hydrogen bomb, promoted by Edward Teller and John Wheeler, and by the Russians under pressure from Joseph Stalin, resulted in successful hydrogen bomb tests. The research that went into developing these hydrogen bombs also further improved physicists' understanding of nuclear fusion. Hans Bethe at Cornell and others were able to use this knowledge finally to prove that nuclear fusion, such as when deuterium and hydrogen atoms fuse to form helium, was the main source of heat and pressure that stabilized stars such as the Sun, as Eddington had suggested more than three decades before.

This knowledge of the nuclear processes in stars prompted physicists Sterling Colgate and Richard White to perform extensive numerical calculations on the evolution of black holes through stellar collapse—calculations that were also significant for static solutions such as the Oppenheimer-Volkoff solution.[7] With the best knowledge of nuclear fusion processes built in to their computer code, they discovered that the pressure produced by the sophisticated nuclear processes in stellar collapse would not prevent the star from collapsing to a black hole when it had a mass greater than three solar masses. This research, which again verified the earlier conclusions of Oppenheimer and Volkoff, finally convinced the physics community that, unless there was some other possible source of pressure in a star, if the star is massive enough, it must collapse to a black hole. Thus, during the 1960s, the physics community began to take the reality of black holes seriously.

HAIRLESS BLACK HOLES

What happens when a star with magnetic fields and other physical characteristics, such as being rather lumpy, collapses to a black hole? This was

---

7. S.A. Colgate and R.H. White, "The Hydrodynamic Behavior of Supernovae Explosions," *Astrophysical Journal*, **143**, 626–681 (1966).

a problem relativists and astrophysicists were concerned about during the 1960s. Vitaly Lazarevich Ginzberg was the first to consider this problem. Ginzberg was famous in Russia for his pioneering work in understanding superconductivity. He authored an article with Lev Landau in a Russian journal that was the first article to investigate the mechanism leading to the near free flow of electrons in metals at temperatures close to 0 Kelvin, which is called *superconductivity.*[8] Ginzburg and Ozernoy pondered the question of what would happen when a star possessing magnetic fields— and most of them do—collapsed to a black hole. He was able to show, but not rigorously, that as the star went through its Schwarzschild circumference, forming an event horizon, the magnetic fields would be forced in through the event horizon and the black hole formed would be perfectly, spherically symmetric.[9]

The problem was also taken up by Zel'dovich and his student Igor Novikov, who contemplated what would happen when a severely deformed star— even, say, a square star—collapsed to a black hole.[10] Would the black hole retain the initial star's deformations? Producing a rigorous mathematical proof one way or another for this problem was exceedingly difficult. However, the Russian physicists were able to solve a simpler problem, in which the star had a small protrusion, such as a mountain, on its surface. Would this protrusion persist when the star collapsed to a black hole?

They used what in mathematics is called *perturbation theory,* for which the main answer is determined by small quantities that, when summed, give the answer to the problem. The perturbation calculation methods for stellar collapse and black holes were independently worked

---

8. V.L. Ginzburg and L.D. Landau, *Zh. Eksp. Teor. Fiz.*, **20**, 1064 (1950). English translation in: L.D. Landau, *Collected Papers*. Oxford: Pergamon Press, p. 546 (1965).

9. V.L. Ginzburg and L.M. Ozernoy, "On Gravitational Collapse of Magnet Stars," *Soviet Physics—JETP*, **20**, 689 (1965).

10. Ya B. Zel'dovich and I.D. Novikov, "Relativistic Astrophysics, Part I," *Uspekhi Fizicheskikh Nauk*, **84**, 877 (1964). English translation in *Soviet Physics—Uspekhi*, **7**, 763 (1965); Ya B. Zel'dovich and I.D. Novikov, "Relativistic Astrophysics, Part II," *Uspekhi Fizicheskikh Nauk*, **86**, 447 (1965). English translation in *Soviet Physics—Uspekhi*, **8**, 522 (1966).

Figure 3.3. John Archibald Wheeler, a giant in the field of gravitation and black holes during the 1960s. He coined the very term *black hole*. Credit: *The New York Times*, 1967, reprinted in obituary article April 14, 2008

out and published by John Wheeler (Figure 3.3) and his student Tullio Regge.[11] They discovered that the star's collapse would produce a black hole with the mountain-like protrusion erased, so that the black hole would form a perfectly spherical shape. In his usual inimitable way, Wheeler described the final, perfectly round black hole as a black hole "with no hair."

Relativity theorists both in the United States and Europe, mainly at Cambridge University under the guidance of Dennis Sciama, tackled the much more difficult problem of proving the no-hair theorem for any size of "mountain" protrusion on the collapsing star, or any size of magnetic field and deformity from spherical symmetry. The final rigorous proofs showed that whatever the distortions of an initially collapsing star, the collapse would produce a black hole with *almost* no hair, the remaining "hair" being its mass, electric charge, and angular momentum. If the idealized initial star had no spin, or angular momentum, the resulting black hole would have a perfectly spherical shape. Dennis Sciama's students were Brandon Carter and Martin Rees in the United Kingdom, and those

11. T. Regge and J. Wheeler, "Stability of a Schwarzschild Singularity," *Physical Review*, **108**, 1063–1069 (1957).

working in the United States and Canada were Richard Price and Werner Israel.

The black hole no-hair theorem postulates that solutions of the Einstein-Maxwell field equations of gravitation and electromagnetism are completely characterized by only three externally observable, classical parameters: mass, electric charge, and angular momentum. All other properties for which "hair" is a metaphor about the matter that formed the black hole disappear behind the black hole event horizon. There is no rigorous mathematical proof of a general no-hair theorem, and it is referred to as the *no-hair conjecture*.

4

Early Observations of Black Holes

What was known observationally about black holes before the current data coming in from LIGO and the Event Horizon Telescope (EHT)? According to general relativity, if one or both stars in a binary star system has a mass too big to form a gravitationally stable collapsed object such as a neutron star, it must form a black hole. Although astrophysicists could not see a black hole directly, clever detective work suggested reliable information about the black hole.

Historically, one way to observe a black hole and determine its mass and angular momentum, or spin, was through observing electromagnetic processes in the form of atomic spectral lines. The X-ray binary systems consist of a progenitor star, which can be a neutron star or a massive star, and a dark companion object. Gas around the progenitor star flows out and around the dark companion, forming a figure eight. Electromagnetic radiation in the form of X-rays emitted from the gas accreting around the dark companion tells us about its nature. Moreover, from the astronomical measurements of the orbits of the progenitor star and its companion, the mass of the dark companion can be determined.

X-RAY BINARIES DISCOVERED

The first such X-ray binary system identified as containing a black hole was discovered by astronomers in 1972. Three of them were Thomas Charles

Bolton, Louise Webster, and Paul Murdin. They discovered an optical star, HDE 226868, orbiting around an optically dark companion, Cyg X-1, which was emitting X-rays. They deduced from astronomical observations, such as the orbital motions or "wobbling" of the companion star, that there was an invisible object in the system. Furthermore, the mass of the invisible star was more than three times the mass of the Sun. Indeed, they found that it was almost six times the mass of the Sun, which meant, according to general relativity, that it must be a black hole. During the more than four decades since then, many such X-ray binary systems have been observed in our Milky Way galaxy.

Electromagnetic observations of X-ray binary black holes, since their discovery, have shown that the mass of a black hole is not more than about 10 times the mass of the Sun. These objects are called *stellar-mass black holes* and must be distinguished from the supermassive black holes that lurk at the centers of galaxies, such as Sagittarius A* at the center of the Milky Way. There exists a gray area in the spectrum of black hole masses between the stellar-mass X-ray binary star of up to about 10 solar masses and the supermassive black holes with masses that are millions of times the mass of the Sun. Such intermediate-size black holes have been concluded to exist because LIGO data appear to show coalescence of binary black holes of more than 40 solar masses.

We have been able to infer information about a binary black hole from observing the physical characteristics of its accretion disk, which is the matter—in the form of gas—that has been sucked from the companion star and circles the black hole (Figure 4.1). Certain radiation signatures in the spectral lines emitted by a black hole's accretion disk can be used to obtain information about the black hole, such as its speed of rotation. The observed spectral lines of the X-rays play an important role in deducing information about X-ray binary black holes.

MONSTERS AT THE CENTERS OF GALAXIES

As for the supermassive black holes, we have excellent evidence that they do exist at the centers of galaxies. In particular, observational evidence

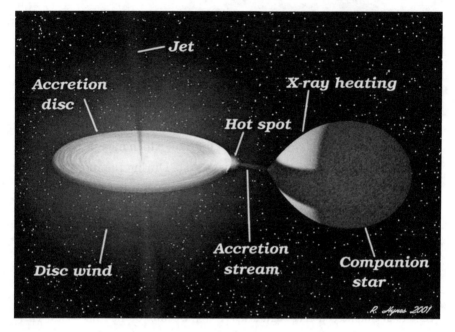

Figure 4.1. X-ray binary. Credit: NASA/R. Hynes

confirms that a supermassive black hole exists at the center of our Milky Way galaxy, and also in other galaxies such as the radio galaxy Centaurus A and M87. These observations are in the form of infrared optical images of the orbits of stars near the black hole, and from millimeter-wavelength radio telescope observations of the physical phenomena associated with the accretion disks surrounding Sagittarius A*, such as radiation emission spectral lines and magnetic fields.

Astronomers have determined that such a supermassive object cannot be an ordinary star, and evidence appears to show that it cannot be a large conglomeration of compact bodies such as neutron stars. Moreover, observations of the gaseous accretion disk surrounding the supermassive body suggest that it does not have a stellar surface, but instead has the event horizon of a massive black hole. However, the evidence for the existence of an event horizon is still controversial and remains to be vindicated observationally.

The supermassive black hole in our galaxy is located about 26,000 light years from Earth in the constellation Sagittarius; the black hole is called *Sagittarius A* (Sgr A* or A-star)*. We know from astronomical observations of the elliptical orbits of stars around this huge black hole that it has a mass of about 4 million times the mass of the Sun. General relativity predicts that such a huge and compact object must be a black hole. The orbits of stars about this black hole are very eccentric elliptical orbits, suggesting the stars are orbiting a supermassive object (Figure 4.2). Just as the mass of the Sun is inferred from the elliptical orbits of the planets, using Newton's law of gravity, so the mass of Sagittarius A* is estimated by observing the stars orbiting it.

Astrophysicists theorize that the supermassive black holes at the centers of galaxies are the sources of enormous energy. The energy, in the form of atomic spectral line emissions, is radiated by large accretion disks surrounding the supermassive black holes. These black holes at the cores of galaxies, combined with the quasars that are sometimes associated with them, are called *active galactic nuclei* (AGNs). They are radiating large amounts of energy. This explains what was initially a puzzle: the mechanism by which quasars produce exceedingly large amounts of energy.

Figure 4.2. Artist's impression of supermassive black hole with a gaseous accretion disk emitting a jet of particles. Credit: Getty Images

QUASARS

Quasi-stellar objects, or *quasars* for short, are among the brightest objects in the universe. Conglomerates of hot gas, they emit enormous amounts of energy, about 10^{41} watts. This is about a thousand times the energy emitted by the entire Milky Way, which consists of 200 to 400 billion stars. The radiation emitted by quasars covers the full electromagnetic spectrum, from X-rays to the infrared, with peaks in the ultraviolet band. Certain quasars emit radio waves and gamma rays. Quasars are among the oldest objects in the universe. Redshift measurements have shown that quasars can be found not long after the Big Bang. It takes billions of years for their light to reach us. According to Hubble's Law, a redshift of receding galaxies and quasars is related to the distance to them. The measured redshift of the receding quasars reveals they are at large cosmological distances from Earth.

The first quasar was discovered in 1963 by astronomer Maarten Schmidt at Caltech, using the Mount Palomar observatory. At that time, it was considered the most distant object ever observed, and it was so bright that it was mistaken for a star, which is why it is called a quasi-stellar object. But it could not be a star, because it was a billion light years away, and an ordinary star would not produce sufficient radiated energy to be seen at such a distance. In addition, astronomers realized that the spectral lines of the object did not coincide with those we know are associated with stars.

Schmidt discovered quasar 3C 273 (Figure 4.3) because he was studying the object as a radio source. It appeared as if the radio signals were coming from a star and he could not figure out what the elements would be that could produce such bright spectral lines. Then he realized the emission lines were from hydrogen gas, but had been shifted significantly to other wavelengths. This shift in the spectral lines could only be explained on the basis of Hubble's Law, which determines the expansion rate of the universe. Thus, object 3C 273 must be located billions of light years away. From this, he concluded that to be as bright as it appeared, it must be brighter than a million galaxies at that great distance.

Figure 4.3. Hubble Space Telescope photo of quasar 3C 273. Credit: ESA/Hubble and NASA

The farthest quasars exist in the early universe, about 800 million years after the Big Bang, and can still be observed as points of light. The brightest quasar is Schmidt's 3C 273 in the constellation Virgo. It is so bright that it can be seen by an amateur's telescope, even though it is so far away. The measured luminosity of quasars varies rapidly in the optical range, and even more rapidly in the X-ray range. From these measurements, we can determine the volume of a quasar, and deduce that quasars are not much bigger than our solar system. The farthest known quasar is ULAS J1120+0641, with a redshift of about seven. This means that its comoving distance is about 29 billion light years from Earth. The comoving distance from Earth to the quasar factors out the expansion of the universe. Two hundred thousand quasars have now been observed by the Sloan Digital Sky Survey (SDSS).[1]

1. The SDSS, an astronomical project centered in New Mexico, aims to produce a full sky survey of galaxies using a 2.5-meter-wide computerized optical telescope and multispectral imaging.

Astrophysicists conjectured early on that a spinning black hole at the center of a quasar is the engine that produces its enormous radiated energy. Quasars are situated near the centers of some galaxies. Ordinary galaxies, such as our Milky Way, which do not contain quasars, do contain central, supermassive black holes with an energy output much less than quasars.

The black holes at the centers of quasars are accreting matter at an enormous rate. The friction caused by this accretion process heats up all the gas around the black hole to produce the enormous flux of radiation energy that we observe. The mass of the central supermassive black hole in a quasar can range from 100,000 solar masses to more than a billion solar masses.

It is speculated that quasars are formed from the collisions of ordinary galaxies, such as our Milky Way and the Andromeda galaxy, and these collisions take 3 to 5 billion years to complete. The age of the universe, of about 14 billion years, can permit many such collisions to occur. During the collision, most of the stars in the colliding galaxies are ejected, and escape. What remains is the coalescence of the two supermassive black holes at the cores of the galaxies. These become the very energetic black holes in the resulting quasars. How quasars form is currently a controversial issue.

Schmidt's discovery of quasars falsified the Bondi-Hoyle-Gold steady-state theory of cosmology. Popular during the 1950s, this theory assumed the universe was uniform in its matter content, and therefore young galaxies should be evenly distributed throughout the universe, some close and some far away. Schmidt's observations, however, showed that young galaxies are *only* far away. These young, active galaxies were so distant in the universe that their existence repudiated the steady-state assumption that all galaxies are evenly distributed.

Another, less luminous, set of AGNs are called *Seyfert galaxies*, which together with quasars are two of the largest groups of active galaxies with a quasar-like nucleus. Most of the Seyfert galaxies look like spiral galaxies, but their cores have a luminosity equivalent to that of the whole Milky Way.

IMPORTANCE OF THE EVENT HORIZON

To establish that a black hole is the object predicted by general relativity, we have to confirm observationally that these massive objects have an event horizon—the spherical surface from which not even light can escape. The early observations of black holes did not prove the existence of such event horizons. Indeed, models have been proposed in which event horizons are not part of the makeup of these massive dark objects. The evidence for the existence of the event horizon is circumstantial, based on controversial theoretical models using information about the accretion disks surrounding collapsed dark objects.

One observation that could provide vital clues about the existence of black holes and event horizons is if the putative black hole were to be seen to swallow matter in the form of stars. However, this is too small an effect to be observed with any significant accuracy. Moreover, we are faced with the issue that distant observers, such as ourselves on Earth, are unable to *see* matter falling into a black hole because of the nature of the coordinate time associated with our reference frame on Earth. Only observers close to or falling into the black hole would be able to detect directly other matter falling into it.

In an article published in 2009, Avery Broderick, Abraham Loeb, and Ramesh Narayan investigated whether the supermassive dark object at the center of the Milky Way is indeed a black hole.[2] By definition, because a black hole has an event horizon, it does not emit any directly observable radiation. (Recall that Hawking radiation is not directly observable.)

The observations used by Broderick, Loeb, and Narayan showed that the dark compact object at the center of our galaxy is radiating virtually nothing. It could only possibly emit less than a maximum of 0.4 percent of the luminosity, or brightness, observed at Sagittarius A*, including its accretion flow. If the object is only producing less than 0.4 percent of the visible light of Sagittarius A*, then the accretion flow of hot gases

2. A. Broderick, A. Loeb, and R. Narayan, "The Event Horizon of Sagittarius A*," *Astrophysical Journal*, **701**, 1357–1366 (2009).

has to produce about 99.6 percent of the light actually observed from the system. The total observed luminosity of the system is surprisingly small, considering the mass of the object. The authors concluded that the object has to be a black hole with an event horizon because, in effect, almost 100 percent of the luminosity is produced by the accretion flow. Otherwise, the very massive dark object would be radiating detectable light. This is a clever and indirect way of inferring the existence of a supermassive black hole.

The authors argue that if the dark object at the center of the galaxy does not have an event horizon but is assumed to have a *hard* surface like a star, then the gas falling in toward the hard surface will have kinetic energy, which is converted into radiation inside or on the surface of the compact dark object. The authors assume the dark object is in thermodynamic equilibrium, which means the amount of radiation emitted and absorbed is equal. However, if it is a black hole with an event horizon, it would absorb 100 percent of incoming radiation and not emit any. The authors claim that if the object is not a black hole, then the amount of emitted radiation should exceed what astronomers observe by 10 to 100 times.

To come to the conclusion that the dark object is a black hole with an event horizon, the authors had to make certain assumptions. A crucial assumption is that a nonblack hole dark object reaches a steady state during its lifetime—meaning, it does not grow or decrease in mass. The argument for the object reaching a steady state is based on the knowledge of how a compact object such as a neutron star evolves in time. Indeed, the supermassive object can evolve over a long time, nearly the age of the universe, so the authors presume it can reach a steady state.

However, a real black hole cannot ever reach a steady state because all matter that falls into it increases its mass. None of the swallowed matter can escape the black hole event horizon, so the black hole cannot be in equilibrium or a steady state. It must continuously grow.

Another significant assumption the authors make is that the alternative to the black hole—the dark compact object—has a hard surface. This does not literally mean the surface is "hard" like a mirror, but that it reflects radiation. This may not in fact be the case, because we do not know what matter could comprise the dark object, and we certainly do not know whether it would have a hard surface.

A more recent article by Wenbin Lu, Pawan Kumar, and Narayan emphasized that without making these assumptions, it is not possible to conclude definitively that the compact dark objects at the centers of galaxies are indeed black holes with event horizons.[3] On the other hand, these authors argue that the assumed properties of the accretion disk, associated with stars colliding with the surface of the dark object and plunging into it, strongly suggest an event horizon exists.

We can draw an analogy with this situation. Picture a theater stage with the curtain closed. We can't see what's behind the curtain. However, the audience can suddenly see that the curtain is rippling and bulging from movement behind it. From this observation, we deduce there are stagehands or actors moving on the stage behind the curtain. Although we can't see them directly, because of the curtain, we have inferred from the behavior of the curtain that they do exist. Similarly, we cannot observe matter falling into the compact dark object at the center of the galaxy, but we can observe the accretion flow outside the object as it exhibits observational phenomena like radiation. We can infer from these phenomena what is going on within the compact dark object. Yet we still do not know whether the object is indeed a dark object or a black hole.

HUNTING FOR EVENT HORIZONS

Despite the difficulties in observationally pinpointing the actual existence of an event horizon, astrophysicists believe they have identified supermassive black holes at the centers of several galaxies, and that they have identified many stellar-size black holes in X-ray binary systems in our galaxy

3. W. Lu, P. Kumar, and R. Narayan, "Stellar Distruption Events Support the Existence of the Black Hole Event Horizon," *Monthly Notices of the Royal Astronomical Society*, **468**, 910–919 (2017).

alone. However, currently there is no incontrovertible proof that these massive dark objects are true black holes with event horizons.

Until recently, the observations of X-ray binary star systems, and the stars and radiation near the supermassive objects at the centers of galaxies, have been the only ways we have had to infer the existence of black holes. With the detection by the LIGO/Virgo collaboration of gravitational radiation, or waves, produced by the collision of two binary black holes, we now have a new observational source of information about black holes. Moreover, the rich observational information that will come from the EHT will also be a dramatic step forward in our understanding of these mysterious objects. One of the astonishing aims of the EHT project—or the Very Long Baseline Interferometry (VLBI) project—is to observe the shadow of the supermassive black hole at the center of our galaxy as it projects onto the black hole's bright accretion disk. This "shadow," or silhouette, is expected to show up in the data as a dark area where there are reduced emissions coming from the accretion disk, or from distant light sources behind the black hole. This kind of eclipse phenomenon is a result of the invisible black hole blocking light trapped by the strong gravitational field of the black hole, as well as blocking the radiation coming from the accretion disk.

The observation of the shadow, as well as the emission of light of the accretion disk, will provide important new information about supermassive black holes, in addition to the information obtained from observing the elliptical orbits of nearby stars. (See Chapter 10 for a full discussion of the EHT project.)

From both the theoretical and observational points of view, the most significant feature of a black hole is its event horizon. The fact that a black hole is as "black" as it *is* is a result of the infinite redshifting of light at the event horizon. It follows that the black hole is not visible to the naked eye or optical telescopes.

Let us suppose that astronauts approaching the black hole event horizon turn on a light signal from their spaceship. Distant observers see that the light signal weakens—is shifted to lower, redder frequencies—and starts disappearing the closer the astronauts get to the black hole. If the astronauts are approaching a true event horizon, and are not adjusting the strength of their light signal, then the disappearance of the light signal is a result of it being infinitely redshifted at the event horizon. From the viewpoint of the distant observers, if the astronauts are actually to reach the event horizon, they must be propelled by an infinite acceleration. This, of course, connects back to the fact that distant observers measure time using their own clock, as opposed to the "proper" time of the clock used by the astronauts falling through the event horizon. If the distant observers could see the astronauts' clock flashing every second, it would appear to be running much slower than their own. The distant observers never actually see the astronauts reach the event horizon, but instead see that the spaceship gets "frozen" before it actually reaches the event horizon. This is a theoretical result due to black holes possessing very strong gravity within relativity theory.

The problem with observationally confirming that a black hole actually has an event horizon is two-fold. The distant observers using a telescope have to make observations close enough to be able to measure properties of the black hole close to the event horizon or Schwarzschild radius of the black hole. This requires extremely fine resolution of the telescope. On the other hand, the black hole has the gaseous accretion disk surrounding it, so observers have to try to disentangle the astrophysical effects of the accretion disk from the physical properties of the event horizon. To overcome this problem, theoretical astrophysicists construct models that simulate the reaction of the accretion disk to the presence of an event horizon. Any kind of theoretical model is based on selected hypotheses, which cannot always be tested experimentally, so any conclusions about the existence of the event horizon are subject to these hypotheses being correct.

For example, the amount of radiation energy emitted by the accretion flow can be affected by the existence of an event horizon or the lack of one. Measuring the amount of light or X-rays emitted by the accretion disk of a black hole, and testing the magnitude of this radiation energy against simulations in which an event horizon exists or doesn't exist, can lead astrophysicists to believe or disbelieve that event horizons exist. Clearly, there is a built-in bias in the results, because of the need for hypothetical assumptions, and for this reason, whether the black hole has a true event horizon is still controversial.

Another aspect of the event horizon is that it is a spacetime membrane boundary and not like the solid material surface of an object such as a star. As you may recall, a solid, stellar-like surface would have a pronounced effect on the physical processes of radiation emission by the accretion disk. A spacetime boundary not made of matter would have a different effect on these accretion disk radiation emissions. The current consensus among astrophysicists is that, from these simulation models of accretion disks, black holes do have an event horizon. Again, the consensus depends on certain currently unverifiable assumptions. Astrophysicists hope the observational results obtained by the EHT will ultimately resolve the event horizon controversy.

RETURNING TO THEORY

If we strictly believe in the classical predictions of general relativity for the collapse of a massive body, then the existence of an event horizon for the final black object is inevitable. The mathematics of the theory dictate this result. The final dark object is described in general relativity by the Schwarzschild solution, or by the Kerr solution when the black hole has a rotational angular momentum, or spin. If a black hole has an electrical charge, then general relativity must be combined with Maxwell's electromagnetic equations, and the charged black hole is described by the Reissner-Nordstrøm solution. This solution produces two event horizons, an interior and an exterior horizon, depending on whether the electric charge exceeds the mass of the black hole, in appropriate physical units.

We do not expect a black hole to possess a significant electric charge that can affect the geometry of spacetime, because the electrical repulsive force (Coulomb force) is 40 orders of magnitude greater than the attractive gravitational force because of the mass of the black hole. This repulsive force would blow the black hole apart instantly. Moreover, electric charge comes in the form of positive and negative particles such as protons and electrons, and as with any astrophysical body, the positive charge is neutralized by the negative charge, whereby the body is overall charge neutral. A collapsing star forming a black hole would instantly discharge its electrical charge. So we do not expect real black holes to be described physically by the Reissner-Nordstrøm solution. If, hypothetically, the electric charge in this solution did exceed the mass, then the charged black hole may not have an event horizon, depending on the size of the electric charge. However, we have already excluded this possibility physically, for any negligible electric charge possessed by the black hole would have little or no effect on the spacetime geometry.

There exist alternative theories of gravity that have Einstein's general relativity as a special limit, and solutions of the field equations of such a theory can describe black holes without a central singularity or an event horizon. These are called *regular black hole solutions*.

With the detection of gravitational waves by the LIGO/Virgo collaboration, and with the accumulation of observational data by the EHT collaboration, we hope that critical features of black holes, such as the existence of event horizons, will be either verified or disproved. Finally, the black hole will rise out of the bedrock of theory to become a physically real and fascinating object in the universe.

5

Wormholes, Time Travel, and Other Exotic Theories

In 1935, Einstein and his assistant Nathan Rosen published a strange and intriguing article in *Physical Review*.[1] It expanded on a paper by Austrian physicist Ludwig Flamm, who, soon after Einstein's publication of his field equations for gravity, and Schwarzschild's publication of his solution to the field equations, showed that if you use the Schwarzschild solution, space-time could be split into two distant, flat pieces connected by a tunnel. In the recent movie *Interstellar*, an astronaut illustrates this idea by bending a piece of paper over and spearing the two pieces with a pen, creating two holes and the tunnel. Einstein and Rosen's spacetime tunnel eventually became known as the Einstein-Rosen bridge. During the 1950s, physicist John Wheeler coined the phrase *wormhole* to describe this theoretical phenomenon. The bending over of spacetime is a way to create a wormhole.

WHY WORMHOLES?

The original motivation for the Einstein-Rosen article was to get rid of the singularities in Einstein's gravity theory. They wanted to describe an

1. A. Einstein and N. Rosen, "The Particle Problem in the General Theory of Relativity," *Physical Review*, **48**, 73 (1935).

elementary particle such as an electron by using a geometric spacetime solution of Einstein's field equations. In other words, they wanted elementary
particles to arise from the mathematical equations of general relativity. This
was a novel approach to understanding subatomic particle physics, in which
the particles were envisioned as manifestations of spacetime geometry itself. The original motivation for the Einstein-Rosen bridge as a description
of subatomic particles has not become part of the modern particle physics
community's interpretation of subatomic particles, but the Einstein-Rosen
bridge has become part of the folklore of black holes.

Einstein had an aversion to singularities in his gravity theory, and devoted
much time during the 1930s and 1940s trying to avoid them. The current interpretation of the wormhole is that it is part of black hole geometry. Joining
two separated regions in spacetime by a wormhole can avoid the formation
of a black hole singularity, where the density of matter is infinite. Of great
interest to writers of science fiction, it would also allow for space travel that
would not be limited by the speed of light, as demanded by special relativity. You could travel from one part of the universe to another very quickly
through a wormhole.

PRACTICAL PROBLEMS WITH WORMHOLES

The gateway to a wormhole can be a black hole. However, this handy
transportation shortcut is more fiction than science. The problem is that
in Einstein gravity, a space explorer cannot traverse a wormhole because the wormhole pulsates in time, pinching off the tunnel and thereby
killing the adventurous space explorer partway through it. To avoid this
theoretical pinching off, Kip Thorne and Michael Morris showed that exotic matter with negative energy could prop up the wormhole and keep
it open.[2] But it is generally agreed in physics that such negative energy/
matter does not exist in macroscopic systems. So if you remain true to

---

2. M.S. Morris and K.S. Thorne, "Wormholes in Spacetime and Their Use for Interstellar
Travel: A Tool for Teaching General Relativity, *American Journal of Physics*, **56**, 395 (1988).

standard principles of physics, then, alas, you cannot use a wormhole to travel from one region of spacetime to another. Neither can you avoid the black hole singularities.

If it were possible to construct a traversable wormhole by new physics, such as an alternative gravitational theory, then it might be possible to travel from our galaxy to a galaxy in a different part of the universe within a short period of time. It might also be possible to construct a traversable wormhole such that you could travel backwards in time, thereby violating causality, as H.G. Wells imagined with his fictional time machine. This is, of course, a mind-blowing idea that has been seized upon, since Wells, by many science fiction writers. Many creative physicists have attempted to construct mathematically stable wormholes.

In constructing a traversable wormhole, one risks violating the principle of causality—namely, that a physical effect must happen after its cause. To physicists, violating this principle is an anathema and leads to the grandfather paradox: by violating causality, one can go back in time and kill one's own grandfather before one is born. Again, this Wellsian situation is fine in fiction, but is generally not accepted by the physics community.

However, physicists have speculated that by using quantum physics, it might be possible to avoid violating causality with a wormhole. Heisenberg's Uncertainty Principle can be invoked to save the day. This principle states you cannot measure the position of a particle and its momentum simultaneously. By using this experimental fact, an astronaut could cheat the causality principle at the quantum level when traversing the wormhole, by avoiding a violation of the conservation of energy, and could therefore go backward in time without getting into trouble with causality. These ideas remain in the realm of speculation.

RESEARCHING WORMHOLES

As outlandish as wormholes may sound—and how could anyone actually "construct" one out in the cosmos for spacetime travel?—they are a popular topic among physicists. Many papers have been published endeavoring to construct a feasible, traversable wormhole with exotic matter or, preferably, without it.

Using my modified, alternative gravity theory, called *scalar–tensor–vector gravity*, also known as *modified gravity* (MOG), I published extensions of black hole solutions in general relativity called *Schwarzschild-MOG* and *Kerr-MOG black holes*. Since the publication of my papers on MOG, there has been an extensive literature on MOG published by other authors. In the MOG theory, besides the spacetime metric of Einstein gravity, which consists of purely attractive gravity acting between bodies, there is also a massive spin-1 vector field producing repulsive *antigravity* between bodies. (The spin-1 vector field is a graviton, which is the gravitational quantum equivalent of a photon.) With the help of this repulsive force, I envisioned a material "strut" in a wormhole connecting two black holes such that the strut was made stable by balancing the attractive gravitational force with the repulsive antigravity force in MOG.[3]

This idea was first proposed by two Austrian physicists, F. Schein and P.C. Aichelburg.[4] They used the repulsion created by an electromagnetic charge, rather than gravitation, to hold the wormhole open. In my MOG theory, the electrical repulsion, which keeps the wormhole in equilibrium, is replaced by repulsive gravity, which is produced by two massive material shells instead of electric charge. These material shells inside the wormhole form a kind of strut and are made of ordinary, electrically neutral matter. The inner strut stabilizes the wormhole so that it does not pinch off as the astronaut attempts to pass through it. The strut itself, in my theory, is composed of nonexotic, positive-energy matter that is kept stable by the MOG antigravity force. There is no need to postulate negative energy.

3. J.W. Moffat, "Black Holes in Modified Gravity (MOG), *European Journal C*, **75**, 175 (2015).

4. F. Schein and P.C. Aichelburg, "Traversible Wormholes in Geometries of Charged Shells," *Physical Review Letters*, **77**, 4130 (1996).

TYPES OF WORMHOLES

There are at least two ways of constructing a traversable wormhole. One is a scenario pictured in the film *Interstellar*, in which the huge black hole called "Gargantua" allows for a wormhole to be produced within it, connecting the black hole to distant parts of the universe (Figure 5.1).

The other scenario is two black holes forming a wormhole between them, as in MOG, also connecting distant parts of the universe. From a distance, the two black holes seem to form one big black hole (Figure 5.2), and the wormhole is a narrow tunnel that cannot be seen by distant observers. So in a sense, the two-black holes scenario is equivalent to the one-black hole scenario, as seen by distant observers.

What happens to hapless astronauts venturing into a wormhole who immediately find themselves at some distant part of the universe? The problem is that the construction of the wormhole can be interpreted as a

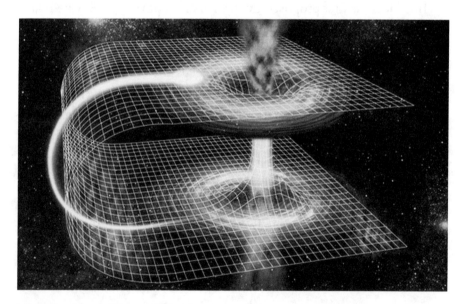

Figure 5.1. Interstellar wormhole. The curvature of spacetime encloses a wormhole connecting two distant parts of spacetime. This was illustrated in the film *Interstellar*. Credit: space.com/royalty-free image from edobric/Shuttercock

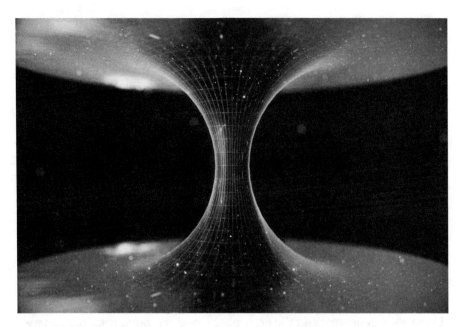

Figure 5.2. Two black holes with a wormhole throat connecting them. Credit: Newsweek 11/1/18, istock

one-way trip in time. To get the astronauts home, we have to interchange the wormhole entrance and exit such that everything is reversed in time. How do we do that? One possible solution is to construct another one of these wormholes not too far away from the first one in space, such that the entry of the wormholes and the exit are reversed in time. This creates a way for astronauts to return to Earth. As in *Interstellar*, the returning astronaut is significantly younger than the colleagues and family left behind on Earth.

There have been many proposals to create traversable wormholes. These include increasing the number of spatial dimensions from the three that we observe, and proposals that speculate on producing exotic matter. None of these proposals based on Einstein gravity has been considered attractive. One problem is to produce a wormhole portal of a macroscopic size that would allow an astronaut to go through. In any event, the idea of a macroscopic traversable wormhole remains in the realm of speculation: it's fun to think about.

USING BLACK HOLES TO DETECT DARK MATTER?

A significant problem in physics and astrophysics today is the need to postulate invisible "dark matter" to explain the stronger gravity observed in galaxies and clusters of galaxies needed to stabilize these systems. In other words, Newtonian and Einstein gravity theories are at odds with observations. See Chapter 9 for a detailed discussion of dark matter.

Roger Penrose proposed in 1971 that one could use a rotating black hole described by Kerr's solution in general relativity to produce the phenomenon called *superradiance*. A rotating Kerr black hole (Figure 5.3) has two concentric event horizons and what is called an *ergosphere*, which is part of the Kerr spacetime geometry.

The word *ergosphere* comes from the Greek word for "work," and an ergosphere is located outside the Kerr black hole's outer event horizon. Using Kerr geometry and the ergosphere, one can mine rotational energy from the black hole, reducing the residual mass and angular momentum, thereby significantly increasing the number of particles orbiting the black hole as corotating and counterrotating particle spheres. This is *black hole superradiance*. A group of theoretical physicists at the Perimeter Institute in Waterloo, Canada, led by Asimina ("Mina") Arvanitaki, has proposed using Penrose's black hole superradiance mechanism to detect dark matter particles.

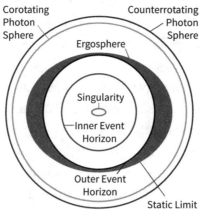

Figure 5.3. Kerr rotating black hole. Credit: Website Universe Review (universe-review. ca/R15-17-relativity04.htm)

Two popular types of dark matter particles that have not yet been detected but have been extensively searched for during the past three decades are the weakly interacting massive particle (WIMP) and the ultralight axion. The Perimeter group claims that for a Kerr black hole spinning with significant angular momentum, a cloud of ultralight axions near the black hole would modify the signature of the gravitational waves produced by the spinning black hole so that the influence of dark matter could be detected by the advanced LIGO/Virgo experiment. One of the ultralight dark matter particles would produce a dominating unstable mode in its wave properties. When combined with the spinning of the black hole—and assuming the black hole is described by Einstein's field equations and has rotational speed close to the speed of light—this could become a significant source of gravitational waves. The gravitational radiation produced in this way can modify the frequencies of the detected gravitational waves compared to those produced without dark matter axions by Einstein's field equations for standard spinning black holes.

One problem with this speculation about using a rotating black hole to detect dark matter particles is that the rotational speed of the black hole has to equal or exceed 80 percent of the speed of light. There is an observational claim that a black hole in the galaxy NGC 1365 spins at a rate of 84 percent of the speed of light. However, from the events now reporting the detection of gravitational waves by the LIGO/Virgo detection networks, we may conclude that stellar binary merging black holes do not have a significant speed of rotation, although for this to become a fact, we need more events to be detected in the future. But the spins of black holes in X-ray binary systems are measured to be a substantial fraction of the speed of light.

HIGHER DIMENSIONS AND GRAVITATIONAL WAVES

Higher dimensions than the three spatial and one time dimension of standard gravitation theory and particle physics have been a dominant research topic for many years. The extra dimensions appear in theories of quantum gravity, such as superstring theories, and the combination of supersymmetric theory and gravitation in what are called *supergravity theories*. One motivation for speculating on extra dimensions is to create models that explain the weakness of gravitation compared to the other forces of nature: the strong force keeping the atomic nucleus together, the weak force responsible for radioactive decay of particles, and the electromagnetic force.

Many of the popular models, such as the Gia Dvali, Gregory Gabadadze, and Massimo Porrati model, published in 2000,[5] and the Lisa Randall and Raman Sundrum model of 1999,[6] postulate a "fifth dimensional bulk" into which gravity can "leak." Some five-dimensional models postulate a four-dimensional "brane." The concept of a brane is borrowed from superstring theory. A string is a one-dimensional object compared to a point particle, which has zero dimensions, whereas a brane is a two-dimensional or higher dimensional object. Mathematically, the four-dimensional brane embraces the standard model of particle physics, whereas the five-dimensional bulk possesses gravitation. There is a warp in spacetime between the brane and the bulk, which can explain the enormous difference in energy between the low energy of the standard model of particle physics, at an energy of about 200 to 500 billion electron volts, and the ultimate Planck energy of about 10^{19} billion electron volts.

Researchers have investigated what happens to gravitational waves in higher dimensions and they have also studied the nature of black holes in higher dimensions. With the advent of the detection of gravitational waves by the advanced LIGO/Virgo project, there now is renewed interest in possibly using gravitational waves to detect the existence of higher dimensions. On the other hand, the Large Hadron Collider (LHC) at CERN in Geneva has been looking for higher dimensions for decades and has found no evidence for them. The fact that LHC experiments have not

5. G. Dvali, G. Gabadadze, and M. Porrati, "4D Gravity on a Brane in 5D Minkowski Space," *Physics Letters*, **B85** (1–3), 208–214 (2000).

6. L. Randall and R. Sundrum, "Large Mass Hierarchy from Small Extra Dimensions," *Physical Review Letters*, **83**, 3370–3373 (1999).

detected supersymmetry particles or the existence of higher dimensions is bad news for superstring theory, which can only be consistently explained as a higher dimensional theory up to 11 dimensions, corresponding to 10 dimension and one time dimension. Superstring theory also requires putative supersymmetric particles as partners to the experimentally known standard particles. Superstring theory can only be mathematically and physically consistent if supersymmetric particles are discovered.

Nonetheless, the negative experimental findings at CERN have not daunted the higher dimensional enthusiasts, who claim that higher dimensions beyond those of the four-dimensional spacetime are needed to meet the goal of unifying the forces of nature, including gravitation, and explaining quantum gravity.

Much effort has gone into understanding how gravitational waves could be emitted from black holes in higher dimensions, as well as how colliding black holes could be a source of gravitational waves in higher dimensions. The higher dimensions will create more physical components in the gravitational waves than those that occur in general relativity in four-dimensional spacetime. The gravitational waves in general relativity in four dimensions have massless modes, like photons and gravitons, and they move at the speed of light. The gravitational waves modes in higher dimensions have what are called *massive modes*, and bring in the famous Kaluza-Klein higher dimensional gravitational theory.

German physicist Theodor Kaluza invented a five-dimensional gravitation theory during the 1920s, and Swedish physicist Oskar Klein later expanded on this theory. In this five-dimensional Kaluza-Klein theory, the higher dimensions produce "towers" of massive particles beyond those known in the standard model of particle physics. Researchers at CERN have attempted to detect these towers of particles by looking for what are called *missing energy events* in the collision of two protons, and have failed to detect them. Nonetheless, the higher dimensional enthusiasts can still believe these towers of particles exist at energies greater than the maximum energy of 14 trillion electron volts (TEV) attainable at the LHC. In the pursuit of understanding higher dimensional gravitational waves, the towers of Kaluza-Klein particles are a source of the extra physical components beyond those in general relativity.

Another possible departure from four-dimensional general relativity is the so-called *warp factor* in the Randall-Sundrum model, which explains the hierarchical difference in energies between low-energy particle physics and the Planck energy, which will also affect the nature of gravitational waves.

In four-dimensional gravitational theory, the gravitational waves are produced by the stretching and squeezing of spacetime. The same stretching and squeezing of spacetime in higher dimensions produces additional components of the gravitational waves compared to those predicted by general relativity. The higher dimensional theories can contain many additional fields beyond the metric field of general relativity, including scalar fields of spin-0 particles. These would produce, besides the two polarization degrees of freedom for the waves in general relativity, additional polarization degrees of freedom resulting from a deformation of the two transverse modes of gravitational waves in four-dimensional spacetime. Gravitational waves, like electromagnetic waves, can involve either longitudinal deformations of spacetime geometry along the direction of motion of the wave or geometric deformations transverse to the direction of motion of the wave (i.e., at right angles to the motion of the waves). Both in general relativity and in Maxwell's theory of electromagnetism, only transverse deformations actually propagate as waves. The additional polarizations of higher dimensional modes of the gravitational wave are called *breathing modes*.

The advanced LIGO experiments and future gravitational wave experiments should be looking for the extra gravitational wave signals—that is, the breathing modes—of higher dimensional waves characterized by a discrete set of higher frequencies determined by the Kaluza-Klein massive modes. It should be stressed that in general relativity and Maxwell's theory, the gravitational and electromagnetic waves are massless waves—that is, the graviton and the photon are considered massless. But in higher dimensional theories, the additional gravitons are massive.

DETECTING HIGHER DIMENSIONAL
GRAVITATIONAL WAVES

The authors of articles on higher dimensional gravitational waves claim these waves occur at very high frequencies. From calculations, they find that the frequency should be as high as 10^{21} Hertz, much higher frequency values than the ones of the recently observed gravitational waves, which are around 150 Hertz. These waves are at much higher frequencies than the upper frequency bound of advanced LIGO, of about 1000 to 10,000 Hertz. Obviously, this does not favor any direct detection of higher dimensional gravitational wave signals of such extremely high frequencies. Indeed, detecting them would require a wholly new type of gravitational wave detection apparatus.

It is a dream of higher dimensional enthusiasts that such a signal, if it were ever available, could be clear enough to serve as a "smoking gun" for the existence of higher dimensions. Unfortunately, one can conceive of other causes for the gravitational wave signal than higher dimensions. Future gravitational wave detector apparatuses such as the Laser Interferometer Space Antenna (LISA) will be sensitive at significantly *lower* frequencies than the current frequencies of advanced LIGO/Virgo. LISA, which will be an interferometer detector in Earth orbit, is designed to detect very low-frequency gravitational waves from colliding super-massive black holes and possibly from primordial gravitational waves from the beginning of the universe. Because of the very low frequency detection of gravitational waves by LISA, these experiments are going in the wrong direction as far as detecting the very high frequency waves associated with higher dimensions.

One of the problems in detecting higher dimensional gravitational waves is that gravitational waves can be very spread out in spacetime. Gravitational waves are emitted continuously for a long time from, for example, black hole binaries getting closer and closer before they collide, which increases the strength or amplitude of the gravitational waves. The detection of these waves only occurs when the amplitude is high enough. In the case of the gravitational waves produced by higher dimensions, we have no knowledge about what produces them, because that depends on unknown physics. However, we can say that the energy of a gravitational wave is proportional to its amplitude times its frequency. It is anticipated that the sources of higher dimensional waves are of low energy. Therefore, the higher dimensional wave signals of high frequency would then require an important energy source, unless the amplitude is low. One can deduce that we need very-high-frequency gravitational waves to compensate for the low amplitude and low energy of the higher dimensional wave signal.

One has to be skeptical about possibly using gravitational waves as a means to detect higher dimensions, regardless of whether you believe in higher dimensions. Detection of higher dimensions through gravitational wave experiments is unlikely because of the extremely high gravitational wave frequencies required. At best, it is a hope that can only be fulfilled in the far future.

BLACK HOLES AS GIANT ATOMS

In general relativity, event horizons of black holes residing in empty space can exist in regions with small spacetime curvatures. This is particularly true of supermassive black holes. For observers outside the horizon, but nearby, such modestly curved spacetime is difficult to distinguish from a spatially flat geometry. But, according to quantum mechanics, black holes, as we have learned, produce Hawking radiation, and the entropy of a black hole is proportional to its horizon area. Entropy is a measure of system organization or disorganization. For macroscopic black holes, these quantum effects cannot be directly measured because of the smallness of the quantities. For example, the temperature of Hawking radiation is inversely proportional to the mass of the black hole, and is too small to be measured for standard black holes. You would need a mini black hole with a mass a small fraction of the mass of the Sun to detect the energy, temperature, and Hawking radiation of the black hole.

Despite these difficulties, a number of authors such as Jacob Bekenstein and Viatcheslav Mukhanov have proposed that quantum effects of black holes can be measured.[7] Bekenstein and Mukhanov suggested that the area A of a black hole horizon is quantized in units of the Planck area. The energy in quantum mechanics, when quantized, means the energy comes in packets proportional to an integer N. The area is also proportional to a dimensionless constant called *alpha*, which is equal to about one. This alpha should not be confused with the parameter alpha in MOG. The area of the horizon can also be written as equal to alpha times the Planck length squared times N. The Planck length is about equal to 10^{-33} centimeters. Because the Planck length is so small, this quantizing of the black hole event horizon area is so tiny we will never observe the Bekenstein and Mukhanov energy packets for black holes. However, the quantized area rule implies that the emission or absorption of radiation by quantum black holes occurs in a series of evenly spaced spectral lines, similar to the case with ordinary atoms. This implies that one can view a black hole as a giant atom emitting and absorbing gravitational waves with well-defined spectral lines.

Bekenstein and Mukhanov explained that the frequency of the radiation can be shown to be inversely proportional to the Schwarzschild radius, which is equal to $2GM/c^2$, where M is the mass of the black hole. In other words, the bigger the black hole, the smaller the frequency of the quantum radiation. You may recall that in quantum mechanics, the energy E is equal to Planck's constant times the frequency of the emitted atomic radiation.

When two black holes coalesce, the final black hole being formed will vibrate like a ringing bell and emit gravitational waves at certain frequencies. The vibrations have an amplitude, or a strength of vibrations, which rings down, oscillating and decaying exponentially over time. This ring-down phase of a perturbed black hole produces what are called *quasi-normal modes of vibration*. The quasi-normal mode frequencies all scale with the

7. J.D. Bekenstein and V.F. Mukhanov, "Spectroscopy of Quantum Black Holes," *Physics Letters B*, **360**, 7 (1996).

inverse gravitational radius GM/c^2 of the black holes, as the radius of the black hole is the only length scale involved in the frequencies. Bekenstein and Mukhanov hypothesized that the frequencies of the vibrations correspond inversely to the wavelength of the gravitational waves.

You may be wondering: will it ever be possible to measure this quantized gravitational wave emission by a black hole behaving like a giant atom?

Decades after the two theorists first thought about these ideas, the advanced LIGO is now detecting gravitational waves. During the first phase of the inspiraling black hole binary system, as detected by LIGO, we can ignore any quantum effects of the event horizons of the two black holes when the black holes are separated by distances much larger than their horizon sizes. When the two black holes collide and coalesce in the final stage of inspiraling, then the black holes merge, resulting in a final black hole with a mass less than or about the size of the sum of the masses of the original two black holes; the final black hole also has an event horizon.[8] The resulting black hole will vibrate and emit gravitational waves, as explained, and the so-called ring-down phase can be modeled by the vibration of a spinning Kerr black hole. The ring-down phase quasi-normal modes have a frequency about the size of the inverse of the gravitational radius, so the frequencies are the same magnitude as quantum Hawking radiation. However, the gravitational waves observed by LIGO in the ring-down phase are theorized to contain a huge number of individual gravitational quanta, the *gravitons*. Gravitons have never been actually detected, so this part of the scenario remains abstract at this time. The gravitons would be produced by quantum transitions between energy levels of the giant black hole/atom separated by quantum units of Hawking gravitational wave temperature. Hawking radiation can be made up of particle–antiparticle pairs, photons, and gravitational waves. So, the quantization idea of Bekenstein and Mukhanov should describe correctly the energy levels of this Hawking radiation. Bekenstein and Mukhanov assumed that the quantum radiation formula

8. For a rotating black hole there are two event horizons, an inner and an outer horizon.

for gravitational waves applies to the highly nonlinear merger phase of the two black holes.

If the Bekenstein-Mukhanov proposal is correct, then the observation of the ring-down phase frequencies should differ from those predicted by classical general relativity. It becomes clear from calculations that it requires many LIGO gravitational wave detections to determine whether the black hole/atom idea of Bekenstein and Mukhanov is correct. The LIGO collaboration expects that future events will have a higher signal-to-noise ratio, permitting a sufficiently precise determination of the ring-down phase. In particular, the measurements of the mass and spin of the final black hole may be determined with sufficient accuracy to allow for a determination of the quantum radiation effects.

It is extremely unlikely that the signal-to-noise ratio for the detected ring-down phase can ever reach the extreme precision required to detect the putative Bekenstein-Mukhanov quantum radiation.

The many speculative ideas that have been spawned by the detection of gravitational waves from merging black holes, and by the intrinsic ideas of black holes and wormholes, reflect the tendency of contemporary physicists to work only with mathematical equations, ignoring whether these ideas can ever be verified or falsified by future experiments.

6

Origins of Gravitational Waves and Detectors

The history of detecting gravitational waves started with Einstein's papers in 1916[1] and 1918,[2] in which he first showed that his gravitational theory predicted the existence of gravitational waves. With the publication of his first paper, he was already subject to criticism. One critic was Italian mathematician Levi Civita, who played an important role in developing the mathematics of Riemannian geometry, which Einstein used to formulate his gravitational field equations.

Civita pointed out that the mathematical expression for the energy of gravitational waves in Einstein's theory was not a generally covariant tensor. This means the physics described by the tensor in the form of energy is not independent of the coordinate system used to observe it. That is, the physics of this gravitational wave energy would look different to observers in different reference frames. One observer could detect the energy of gravitational waves, whereas another observer in another reference frame would observe no energy and, therefore, for this observer, gravitational waves would not exist. This phenomenon challenged the

---

1. A. Einstein, "Näherungsweise Integration der Feldgleichungen der Gravitation," *Preussische Akademie der Wissenshaften, Sitzungsberichte*, (part 1), 688–696 (1916).

2. A. Einstein, "Gravitationswellen," *Preussische Akademie der Wissenshaften, Sitzungsberichte*, (part 1), 154–167 (1918).

basic physical principle of general relativity: that the laws of physics are independent of the choice of reference frames.

In his second paper on gravitational waves, Einstein added an appendix in which he attempted to defend his arguments for the existence of gravitational waves. But later in life, he changed his mind again and negated this defense.[3]

BELIEVING OR NOT BELIEVING
IN GRAVITATIONAL WAVES

During the 1950s and 1960s, there were heated debates among physicists about whether gravitational waves exist. This debate was compounded by the claim, originally made by Einstein, that it would be impossible to detect gravitational waves because of the weakness of gravity.

A leading proponent of the idea that gravitational waves do not exist was Polish physicist Leopold Infeld. Infeld was one of Einstein's important collaborators during the 1930s and 1940s, so his opinions were taken seriously. Another important collaborator was Nathan Rosen, who worked with Einstein and Infeld on the problem of motion in general relativity, and with Einstein on gravitational waves and, as you may recall, wormholes. Rosen, also, doubted the existence of gravitational waves for his entire life.

A proponent of the opposite camp, that gravitational waves do exist, was Vladimir Fock, a Russian physicist who published an important book on Einstein's gravitational theory. His book was called *Theory of Space, Time and Gravitation* and was published in English in 1959.

Fock claimed that the principle of general covariance (i.e., that physical phenomena can be observed independently of the frame of reference) should not be the only criterion for understanding general relativity. He also reexpressed Einstein's field equations as wave equations by choosing to solve them in a special coordinate frame called the *de Donder coordinates*,

3. See Daniel Kennefick's *Traveling at the Speed of Thought* for a detailed historical account of the history of gravitational waves.

after Belgian mathematical physicist Théophile de Donder. These coordinates were also called *harmonic coordinates*, and they originated with the works of Simon Laplace and others when solving Newton's equations of gravity. In these particular coordinates, Einstein's field equations take on the form of wave equations—namely, they describe gravitational waves—and this fact motivated Fock to claim that gravitational waves must exist in nature.

The relativity community at the time did not agree with Fock's dismissal of the basic principle of Einstein's gravitational theory, that the physical predictions of the theory are independent of the choice of coordinates. Fock was a fervent supporter of Communism and Lenin's doctrines, which were critical of Einstein's theory of gravitation. The Soviets claimed that the basic principles of Einstein's gravity theory were tainted by western capitalist ideals.

An important turning point was reached in 1957 at a conference in Chapel Hill, North Carolina, where the topic of the existence of gravitational waves was heatedly debated. Among the physicists actively involved in gravitational research at that time were Herman Bondi and Felix Pirani at King's College, London. Other prominent physicists present at the conference who also believed in gravitational waves were John Wheeler, Richard Feynman, and Joseph Weber. Pirani pointed out an important connection between Newton's second law associated with gravitational tidal forces and the basic geodesic equation of motion of bodies in general relativity. The geodesic equation describes the shortest and straightest path between two spacetime events. Pirani explained that the relative accelerations of two neighboring particles would produce gravitational waves. Technically, this is called *geodesic deviation*, which is the general relativity counterpart of tidal forces in Newtonian gravity.

Feynman objected to the mathematical discussions at the conference between those who believed in gravitational waves and those who didn't. He claimed that the physics of gravitational waves was straightforward, and he promulgated his famous "sticky beads" argument. He said that if you have a rod and you have beads sticking onto the rod, moving backward and forward, then they would sense gravitational waves. The beads moving back and forth on the rod created heat, and the accompanying energy and work demonstrated that gravitational waves had to exist and carry localizable energy.

At the conference, Herman Bondi proposed a similar argument for why gravitational waves must exist and contain energy. This argument also took the form of a sticky beads proposal. Feynman arrived at the conference the day after Bondi gave his talk and registered himself as "Professor Smith." The story goes that when he arrived at the airport, he didn't remember the location of the conference. He asked a taxi driver whether he had transported someone who kept muttering "g-mu-nu-g-mu-nu." This vocalization, $g_{\mu\nu}$, refers to the metric tensor of Einstein gravity. The driver said, "Ah yes! I know where you should be going. I'm taking you to the University of North Carolina."

Feynman and Bondi had come up with the same idea independently, as happens frequently in science. As to registering as Prof. Smith, Feynman likely did not wish it to be known that he had attended this conference on gravity. At an earlier conference, held in Warsaw, Feynman had written a letter to his wife saying he would never attend another conference on gravity because he thought that all these relativists were crazy.

DO GRAVITATIONAL WAVES CARRY ENERGY?

Most skeptics believed in the existence of gravitational waves. Their skepticism arose in how to interpret them. Gravitational waves had already been predicted as early as 1876 by William Clifford of Washington University, St. Louis. He proposed there would be curvature waves:

I hold that (1) small portions of space are in fact of a nature analogous to little hills on a surface, which is on average flat; namely that the ordinary laws of geometry are not valid for them. (2) That the property of being curved or distorted is being continually passed on from one portion of space to another after the manner of a wave.[4]

Clifford envisioned a unified field theory where reality could be described by propagating waves of curvature of three-dimensional space.

Later, others began thinking about gravitational waves. French mathematician Henri Poincaré used the expression *gravitational wave* when speculating about relativistic gravity. Around 1905, when Einstein discovered special relativity, Poincaré suggested, independently, that a relativistic form of gravity would have gravitational waves of acceleration—namely, that a retarded attractive force between two massive bodies would have wave-like propagation at the speed of light. A "retarded attractive force" stands in contrast to the Newtonian gravitational force, which is transmitted instantaneously between bodies. On the other hand, we expect the influence of one body on another to travel at the finite speed of light.

In 1913, Finnish physicist Gunnar Nordstrøm proposed a Lorentz invariant scalar field gravity theory, which had gravitational waves propagating at the speed of light. This proposal did not describe gravity within a geometric scheme, like Einstein's theory of gravity based on Riemannian geometry. Instead, it assumed that spacetime was Euclidian, and only the coordinate transformations of special relativity—namely, Lorentz transformations—were important for gravity.[5]

With the exception of some physicists such as Einstein's assistants Leopold Infeld and Nathan Rosen, most physicists of the 1950s and 1960s believed in the existence of gravitational waves, but they were skeptical about whether these gravitational waves carried energy. The problem was that an analogy was drawn between electromagnetic radiation and

4. W.K. Clifford, "On the Space Theory of Matter," *Proceedings of the Cambridge Philosophical Society*, **2**, 157–158 (1876).

5. For a detailed history and discussion of gravitation theories, see my book *Reinventing Gravity: A Physicist Goes Beyond Einstein* (2008).

gravitational wave radiation. According to Maxwell's equations, without currents and charges, electromagnetic waves propagate at the speed of light, and the energy of the electromagnetic waves is described by a tensor that is invariant under coordinate transformations. Therefore, every observer detects the energy of electromagnetic waves; this energy cannot be transformed away by changing one's coordinates.

The exact opposite is true in Einstein's theory of gravitation: the energy tensor of gravitational waves, as Einstein proposed, is not invariant under the change of coordinates from one reference frame to another. The underlying source of this circumstance in general relativity is the equivalence principle. It states that bodies fall in empty space at an equal rate, independent of their composition. That is, bodies are in freefall in a gravitational field in the absence of other external forces. Many accurate tests have shown that the equivalence principle is experimentally correct. In general relativity, according to this principle, particles move along geodesics—the straightest paths—and in a sense they obey an inertial form of motion in freefall, and do not experience an acceleration, as in Newtonian mechanics. The gravity in Newtonian mechanics is described by a forcefield, a concept that has to be given up in Einstein's general relativity, in favor of the concept of spacetime, which is curved by matter and energy.

However, there is a method called the *post-Newtonian expansion method* of Einstein's field equations that is used by relativists to solve the complicated, nonlinear field equations of Einstein's general relativity. This is an approximation method in which the attractive force of gravity between two bodies is initially assumed to be Newtonian gravity. Then, general relativistic corrections are added for higher and higher orders dependent on the speed of the bodies divided by the speed of light. But this is a particular approximation method, whereas the general interpretation of Einstein's theory is geometric, and gravity between bodies is not a force as in Newton's theory.

Despite the fact that Herman Bondi believed gravitational waves should exist, he was critical of the standard interpretation of them. Bondi pointed out that, according to general relativity, two bodies in a binary orbit both move along geodesics and therefore do not experience an acceleration as in Newtonian mechanics. In Newton's theory, a body is accelerated by gravity. According to general relativity, however, bodies are not accelerated. Rather, they move along a geodesic with inertial motion, which is freefall, and satisfy the equivalence principle. The particles that make up the bodies also move along geodesics[6] and they also do not suffer acceleration. Therefore, the analogy with electromagnetism fails because electrically charged particles, when accelerated, produce electromagnetic waves, which carry energy. Moreover, charged particles do not satisfy the equivalence principle.

How can bodies that do not accelerate, but experience freefall inertial motion, produce gravitational waves? To answer this question, Felix Pirani, at the Chapel Hill conference, promoted his idea that the equations of geodesic deviation corresponding to tidal forces in Newtonian gravity involve the curvature of spacetime. Pirani's equation of motion involved a forcefield described by the curvature of spacetime. The issue of whether gravitational waves propagate with a localizable energy is still controversial today, 60 years after the famous Chapel Hill conference. For example, Canadian physicist Fred Cooperstock and others have published papers demonstrating that gravitational waves do not carry energy.

But if gravitational waves do not carry energy, what is LIGO detecting? An answer is that gravitational waves are a distortion of spacetime, and an interferometry experiment can detect the strain that is caused by a displacement of the interferometer arm, $\Delta L/L$, without having to be concerned about whether the gravitational wave carries energy. For some years, the physics relativity community has promoted the idea that gravitational energy is nonlocal. That is, you cannot measure the gravitational energy in a localized region in spacetime, which amounts to saying that it is not possible to detect gravitational wave energy.

---

6. This may not be true due to dynamical self-forces among the particles.

JOE WEBER AND HIS FAMOUS BAR

The conference at the University of North Carolina inspired Joe Weber to embark upon a long career with the goal of detecting gravitational waves. He is famous for inventing the Weber bar, which was designed to produce a resonant vibration when hit by a gravitational wave, a quite different experimental setup from the LIGO interferometer.

The attempts to measure the minuscule curvature ripples of space-time, which show themselves as gravitational waves, have a long history, going back to the late 1950s and early 1960s. In 1969, Weber claimed to have detected gravitational waves with his Weber bar. This claim was met with loud criticism and inspired much controversy. Weber's fame for discovering gravitational waves was short-lived, as papers were published showing that, because of the weakness of gravitational wave sources, this resonant bar could not possibly have detected a gravitational wave. The resonance that the Weber bar detected was eventually shown to have been caused by Earth-based noise, such as a seismic wave or other environmental disturbances.

Joseph Weber was a lonely pioneer who took upon himself to do what most people, including Einstein, considered impossible—to detect gravitational waves. Weber was born on May 17, 1919, and was a graduate of the U.S. Naval Academy in 1940, with a bachelor's degree in engineering. He continued his naval career by serving in World War II on the aircraft carrier *Lexington*. During the Battle of the Coral Sea, the aircraft carrier was sunk. Having survived this, Weber became a commanding officer of a submarine chaser. Weber's mastery of radar technology and radio was so impressive, that in 1948 he was offered the position of full professor of electrical engineering at the University of Maryland. Remarkable was the fact that he was only 29 years old, and only had a bachelor's degree in engineering.

While he was teaching electrical engineering at the University of Maryland, Weber decided to complete a PhD in physics at Catholic University, supervised by Karl Herzfeld. In the early 1950s, Weber became well known for his work on the physics of lasers, but he never built a laser prototype because of a lack of resources. Unfortunately, although he was nominated, he did not share the Nobel Prize won by Charles Townes, Nikolai Basov, and Aleksandr Prokharov for the discovery of lasers in 1964.

Because of his disappointing research in lasers, Weber decided to seek an avenue of research that was not so competitive. He began to investigate experiments in gravitation theory, and in particular, experiments that tested Einstein's general relativity. He joined John Wheeler's relativity group at Princeton and, during this period, he pursued the possibility of detecting gravitational waves, which was a daunting and isolating choice of research. Like Christopher Columbus, he embarked on a search for a new world, an adventure into the unknown.

During the late 1950s, Weber attempted to produce a gravitational wave detector in every way he could conceive. Finally, he came upon the idea that if the ends of a metal cylinder vibrated ever so slightly as a gravitational wave passed through Earth, these vibrations could be detected and they would signal that a gravitational wave had arrived from deep outer space.

This led to the idea of the Weber bar. The distant acceleration of some massive astrophysical body would produce gravitational waves in the form of a tidal force in spacetime, and would compress and expand the ends of this bar. The bar would have a natural frequency, and when the gravitational wave met this natural frequency of the bar, the ends of the bar would ring, much like a bell or a wine glass.

To realize his idea of a vibrating bar, Weber had to invent a sensor that was sufficiently sensitive to measure the tiny bar-end vibrations. He came up with the idea of using the piezoelectric effect, in which certain crystals and ceramics, when squeezed, in this case by a gravitational wave, would develop electrical voltages from one end of the material to the other. He also had to estimate the frequency of the putative gravitational waves, and he decided on a frequency band of less than 10,000 Hertz (10,000 cycles per second). He chose the resonant frequency of his bars to be about 1000 Hertz or less. This band was in the audible range. As it turns out, this frequency of the gravitational waves was a fortunate guess at this very early stage of experimentation. We now know that the range of the advanced LIGO detection of gravitational waves from the binary black hole collisions is between 10 and 450 Hertz. The frequency of the gravitational waves has to have the same frequency as the natural frequency of the Weber bar in order for the gravitational waves to be detected.

The criticism of Weber's claim that he had detected gravitational waves with his Weber bar was correct. The term *strain* refers to the expansion and contraction of spacetime caused by a gravitational wave originating from deep space. The strain corresponding to the deviation of the LIGO laser beam $\Delta L/L$, where L is the length of the beam and ΔL is its deviation, is about 10^{-21} for the advanced LIGO gravitational wave detections[7]: or one one-thousandth the diameter of a proton. The strain that could be measured by Weber's bar was orders of magnitude bigger than the LIGO strain detection, and so his apparatus could never have detected gravitational waves.

Later, Weber came upon an alternative way of detecting gravitational waves, using the idea of a Michelson-Morley interferometer with lasers to produce light beams along the interferometer arms. Indeed, he was one of the first to propose the idea that later became LIGO.

I first met Joe Weber (Figure 6.1) when I was a research physicist at the Research Institute for Advanced Studies (RIAS) in Baltimore, Maryland during the early 1960s. RIAS was a privately funded physics and mathematics research institute that no longer exists. Because I had studied relativity at Cambridge University for my PhD, I was part of the relativity community at that time. Joe invited me down to the University of Maryland to discuss his gravitational wave detection experiments. I found Joe Weber to be very cordial, yet with an intense demeanor. In his mid forties, he had a shock of steely dark-gray hair, and habitually wore dark-rimmed glasses and a white shirt open at the neck.

Joe took me into an unprepossessing room in the physics department to show me his gravitational wave detector instrument. There was a moment of

---

7. The deviation ΔL equals 10^{-16} centimeters, and the length of the beam is 4 kilometers, or 4×10^5 centimeters. So, the ratio $\Delta L/L$ is about 10^{-21}.

Figure 6.1. Joe Weber with his gravitational wave resonating-bar apparatus.
Credit: University of Maryland libraries

excitement when he led me to his apparatus, which consisted of a 2-meter-long aluminum cylinder covered with little bars of piezoelectric crystals glued around the middle of the cylinder. He had strung the crystals together, forming an electric circuit so that their minuscule oscillating voltages would add up to a strong enough voltage to be detected electronically, an effect that could detect a distance displacement of approximately one tenth or less of the nucleus of an atom—namely, about 10^{-14} centimeters.

I was not, at the time, expert enough in the detection of gravitational waves to appreciate what the magnitude of the spacetime curvature strain of a gravitational wave would be, which could be detected by his apparatus. It all seemed very impressive to me, as a theoretician, and I expressed my admiration for his efforts.

At a relativity conference in 1969, Weber made his announcement that he had detected gravitational waves using his vibrational bar. This was a momentous discovery, and his picture appeared on the cover of *Time* magazine, beginning for Joe Weber a short-lived period of fame.

His "discovery" motivated other physicists to join in the game and develop even more accurate Weber bars. One was Vladimir Braginsky at

Moscow University. He was the first experimental physicist to embark on the detection of gravitational waves after Weber. Braginsky was also, at the time, performing experiments to detect quarks, the fractionally charged particles that are the constituents of protons, neutrons, and other elementary particles. He was also pursuing experiments to verify the equivalence principle, a cornerstone of Einstein's general relativity, which you may recall states that bodies fall in a gravitational field independent of their composition.

Eventually, as other experimentalists built accurate Weber bars, they discovered they were unable to replicate Weber's discovery of a gravitational wave signal. This led to a rather acrimonious period in which Weber had to face mounting criticism. Despite the fierce criticism from his peers, and the mounting consensus that Weber had not detected gravitational waves, he never gave up the claim that he had.

There was a great deal of theoretical activity during the 1970s aimed at determining the magnitude of the strain, or the amount of spatial displacement, needed in a detector to verify the arrival of a gravitational wave from such possible origins as coalescing black holes in a binary black hole system. It turned out that the most accurate Weber bars could only detect a spacetime ripple strain of 10^{-17}. This strain, we now know, was 10,000 times too large to allow the Weber bars to detect gravitational waves emitted from a strong, but very distant, source like the coalescence of black holes.

At an international scientific congress held in China during the late 1980s, which I attended, the conference participants were housed in the former French embassy in Shanghai, where in 1972 Nixon had famously signed an agreement for an open U.S.–China policy. In those days, I was an avid jogger, before jogging became popular. Over breakfast one morning, Joe and I agreed to go out jogging together. So we set out at the break of dawn every morning after that and jogged through the still, dark streets of Shanghai, where throngs of Chinese were practicing tai chi. Even at that early hour, the weather was very humid and we soon broke out into a sweat. At this time, Joe was in his mid 60s and I was in my early 50s. Joe was, as always, lean and healthy looking. During periods of catching our breath, overlooking Shanghai's busy harbor, we would snatch bits of physics conversation, and I talked to him about his gravitational wave experiments.

He was bitter about his treatment by the physics community, and still insisted that he was right in his claim of having detected gravitational waves. I considered it a travesty that the physics community had excoriated this pioneer of gravitational wave detection, although it was unfortunate that Joe would not admit to the impossibility of his experimental attempts actually to detect gravitational waves from some explosive astrophysical source. It is very difficult for a scientist who has devoted so much of his life to succeeding in experimental physics to admit to failure. It is often said that experimental physicists cannot allow for failure even once, whereas theorists—with their fanciful speculations—are permitted to make mistakes all the time without injuriously affecting their careers.

FIRST IDEAS ABOUT LIGO

After the influential 1957 conference at Chapel Hill, a young professor from the Massachusetts Institute of Technology, Rainer Weiss, picked up Felix Pirani's suggestion that gravitational waves could be detected by using light signals. These light signals would detect the ripples in spacetime—or gravitational waves—caused by accelerating massive bodies. Weiss proposed that to detect gravitational waves, one should use a Michelson-Morley interferometer setup, which Joe Weber had also suggested earlier.

In 1887, Albert Michelson and Edward Morley, researchers at Case Western Reserve in Cleveland, Ohio, compared the speed of light moving in perpendicular directions, thereby attempting to measure the relative motion of matter through the ether as Earth moved in its orbit around the Sun. The ether was a purported medium that would carry the electromagnetic waves predicted by James Clerk Maxwell. The scientific community at the time believed in the existence of the ether, although it had not actually been detected.

The Michelson-Morley interferometer (Figure 6.2) consists of a light source, a half-silvered glass plate, two mirrors, and a telescope. The mirrors are placed at right angles to each other, and at an equal distance from the glass plate, which is oriented at an angle of 45 degrees relative to the two mirrors.

Michelson and Morley were looking for interference fringes between the light that had passed through the two perpendicular arms of the apparatus. This would happen because the speed of light would be faster moving along an arm that was oriented in the same direction in which the ether was moving, and slower if oriented in the perpendicular direction. If there was a difference in the speed of light, the crests and troughs of the light waves in the two arms would arrive and interfere a little out of sync, producing a small decrease of intensity of light. However, the experimenters found no discernible fringes indicating a different speed in any orientation, or for any position of Earth in its annual orbit around the Sun.

Indeed, the difference in the speed of light expected by the experimenters was found not to exist, and therefore the ether was deemed not to exist. This result initiated a line of research that led to special relativity, a theory that is free of the ether. The Michelson-Morley experiments have been repeated many times, with steadily increasing sensitivity. Recent optical resonator experiments confirmed the nonexistence of the ether at the accuracy level of one part in 10^{17}. Other types of experiments, such as the Ives-Stillwell and Kennedy-Thorndyke experiments, performed the fundamental tests of relativity theory, in which the speed of light is constant

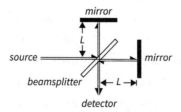

Figure 6.2. Diagram of the Michelson-Morley interferometer. Credit: Scienceworld. wolfram.com

with respect to all inertial frames. That is, the speed of light is independent of the motion of the observer.

Famous Dutch physicist Hendrik Lorentz and Irish physicist George FitzGerald tried to save the ether by concocting the idea that all objects moving relative to a motionless ether underwent a contraction in their length in the direction of their motion. This would mean that the Michelson-Morley interferometer would be unable to detect the ether. This had the unpleasant consequence that the Lorentz-FitzGerald ether theory could not be falsified, and it competed with the new theory of special relativity, on which Einstein and French mathematician Henri Poincaré were working. However, the simplicity of special relativity, and the verifying experiments that followed, prevailed over the Lorentz-FitzGerald theory.

Thus, the ether was finally laid to rest. But, the device that recorded its demise lives on more than a hundred years later. The Michelson-Morley interferometer gave birth to the LIGO interferometer, which can measure phase differences in fringe patterns, much as Michelson and Morley wanted to do in their original ether detection experiment.

WHERE CAN GRAVITATIONAL WAVES COME FROM?

The first gravitational waves detected by the advanced LIGO experiment were emanating from binary black holes that were coalescing—merging into one. The violent event produced ripples of strong gravitational waves in spacetime, like a large rock thrown into a pond. Another source of gravitational waves that has been detected by LIGO is the merging of binary neutron stars. Are there other objects and events out in space than can produce gravitational waves that we can detect on Earth?

The merging binary black holes (Figure 6.3) that are the source of the gravitational wave detection events at LIGO are stellar-mass black holes, with binary component masses up to more than 40 solar masses. The estimated component masses in the range of about 10 to 50 solar masses for the events detected by the LIGO/Virgo observatories were not anticipated by astrophysicists, who had estimated the amplitude of gravitational

Figure 6.3. Artist's depiction of two merging black holes. Credit: NASA

waves from merging black holes. They anticipated the masses would be about 10 solar masses for the black holes, as deduced from observations of X-ray binaries in our galaxy. This has led to what is called the *intermediate stellar-mass black hole conundrum*. What could give rise to black holes of such masses? A decade of investigation of the evolution of black holes would appear to not permit such massive black holes to exist. To create them, the progenitor stars would have to be very massive, the amount of heavy metals in the stars negligible, and the loss of mass during the formation of the black hole also negligible. This has led to a controversy, which is still going on, about the permitted sizes of intermediate stellar-mass black holes.

These stellar-mass binary black holes should be distinguished from the more speculative supermassive black hole binaries, which would be formed by the collision of two galaxies containing central black holes. The latter would only be observed as gravitational wave sources at very low frequencies, or about 10^{-9} Hertz, which would require a huge interferometer in space, such as LISA and the Chinese satellite project TianQin. It was William Fowler who, during the early 1960s, came up with the idea of colliding quasars producing supermassive black hole binaries. Fowler was influential. He went on to win the Nobel Prize in 1983, with Chandrasekhar, for detecting the carbon cycle resonance. This resonance was predicted by Fred Hoyle and was important for the understanding of the beginning of life in the universe.

There are other ways for the universe to generate gravitational waves, aside from two coalescing black holes of whatever size. The more massive and dense the objects that orbit one another, the stronger the signal of the gravitational waves. The speed at which Earth moves around the Sun is too slow to create enough acceleration of masses ever to produce a detectable gravitational wave.

A significant breakthrough by the LIGO/Virgo detection network was the discovery, on August 17, 2017, of gravitational waves spawned by the merging of two neutron stars, and the simultaneous optical discovery of the source: a short gamma ray burst (GRB). The neutron stars in a binary system can be formed by the collapse of two massive progenitor stars orbiting each other; both eventually spiral in and coalesce to form a final neutron star or a black hole, depending on the initial masses of the stars. The final merging can produce a strong enough gravitational wave signal to be detected by the LIGO/Virgo project and future gravitational wave experiments.

How can one tell whether a gravitational wave signal emanates from merging black holes or merging neutron stars? This is done by analyzing the gravitational wave data and determining the masses and spins of the coalescing objects.

An important aspect of neutron star binaries is that neutron stars have already been observed, whereas we have not directly observed isolated black holes. We have observed X-ray black hole binaries, but these black holes with their accretion disks are associated with companion stars.

In 1974, a binary neutron star system was discovered by Joseph Taylor and Russell Hulse using the large Arecibo radio telescope in Puerto Rico. The neutron stars were giving rise to pulses of light, and therefore are called *pulsars*. Neutron stars are formed as the core remnants of exploding supernovae. An example is the neutron star at the center of the Crab Nebula, which is the remnant of a supernova explosion first seen by the Chinese in 1054. A neutron star is only 11 to 40 kilometers in radius, with a mass of 0.9 to 2 times the mass of the Sun. The neutron star density is so high that one teaspoonful of neutron star matter weighs about 10 million tons.

Recall that neutron stars can also form from massive progenitor stars that collapse. Sometimes the neutron stars come from a binary star system. As two neutron stars in a binary neutron star system orbit around each other, the system emits gravitational waves, and the gravitational wave loss of energy causes the distance between the neutron stars to decrease at a measurable rate. Hulse and Taylor measured this "orbital decay" with considerable accuracy, and the value of the decrease agrees remarkably well with the prediction of general relativity for gravitational waves. This discovery was the first indirect evidence that gravitational waves exist, and Taylor and Hulse won the Nobel Prize in 1993 for discovering this binary pulsar.

Yet another source of gravitational waves can be the inspiraling of a black hole and a neutron star, with their eventual coalescence. These sources are also on the list of possible astrophysical events that the LIGO/Virgo collaboration is searching for, and such an event has in fact been detected.

Sources of gravitational waves, then, can be binary black hole systems, binary supermassive black holes, binary neutron stars, black hole/neutron star binaries—and perhaps even the origin of the universe, the Big Bang itself.

ECHOES OF THE BIG BANG

One of the great successes of modern physics is the standard model of cosmology. Based upon Friedmann's solutions to Einstein's field equations in general relativity for the large-scale structure of the universe, we are able to obtain a picture of the early universe up until the time of the formation of stars and galaxies 300 to 400 million years after the Big Bang. Physicists postulate that dark matter makes up about 25 percent of the total energy-matter content of the universe, whereas visible matter comprises only about 5 percent. The remaining 70 percent is supplied by assuming that there exists a dark energy repulsive force causing the observed acceleration of the expansion of the universe. This so-called *concordance model* can explain the considerable wealth of early universe cosmology data now available thanks to large telescope and satellite observations.

In particular, the model can describe the cosmic microwave background (CMB), which is the "afterglow" of the Big Bang explosion about 380,000 years later. The CMB was discovered by Arno Penzias and Robert Wilson as a uniform "noise" observed unexpectedly by the radio telescope at Bell Laboratories in New Jersey in 1964. Further measurements of the CMB have been taken by spacecraft experiments such as the Wilkinson Microwave Anisotropy Probe (WMAP) and the Planck satellite. The standard model of cosmology is able to describe successfully the uniform temperature of the CMB—the so-called *Planck temperature*, at about 2.7 Kelvin—with remarkable accuracy.

The early universe at the Big Bang, and up until about 380,000 years afterward, consisted of a hot plasma of electrons and protons interacting with photons. Because the temperature decreases as the universe expands, it reaches a critical temperature at the so-called *surface of last scattering*, when the photons no longer interact with electrons and then the hot plasma ceases to prevent light, or photons, from escaping and streaming out toward us in the current universe, where it can be observed by telescopes and spacecraft experiments. Eventually, the free electrons and protons combine to form hydrogen atoms or hydrogen gas. Then, later, they re-ionize, and a fraction of the electrons become free of the protons and allow for heavier atoms and for structures such as stars and galaxies to be formed. This stage in the evolution of the universe is called the *period of recombination*, or, metaphorically, the *dawn of the universe*.

But not all the light at the surface of last scattering escapes. The photons that remain in the plasma produce an opaque veil that prevents us from "seeing" the Big Bang. However, the primordial gravitational waves coming from the Big Bang are not blocked or affected by the plasma. If LIGO or perhaps the much larger LISA in space is able to detect the primordial waves, we will finally have a much better picture of how the universe began about 14 billion years ago.

BICEP2, which stands for "Background Imaging of Cosmic Extragalactic Polarization," was an international collaboration at the South Pole to attempt to observe primordial waves emanating from the Big Bang. In March 2014, the group claimed that the observations of the Keck Array telescope at the South Pole had discovered primordial "B-mode" polarization of the cosmic microwave background. These B-mode polarizations were understood to be a unique signature of primordial gravitational waves. However, the results were thrown into doubt when the European Space Agency's Planck collaboration discovered that the BICEP2 polarization results were, in fact, a foreground effect caused by dust within the Milky Way. The BICEP2 collaboration then withdrew their claim to have found evidence for primordial gravitational waves.

The Planck collaboration maps of the foreground dust, released in September 2014, showed that the polarized emission from the dust is much more significant over the entire sky than BICEP2 had allowed for. Further analysis of the data from both BICEP2 and the Planck collaboration put the issue to rest when they clearly showed that the dust produced the main polarization data. This excluded the possibility of having observed primordial gravitational waves. Attempts are now underway to separate observationally the foreground dust B-mode polarization from a polarization caused by primordial gravitational waves.

With LIGO, the Planck satellite observations, and other technologically sophisticated projects, we have come a long way from the early days of gravitational wave detection with Joe Weber and his resonating bar.

7

The Biggest Ears in the Sky: LIGO

It is the morning of February 11, 2016. Rumors have been swirling around the Perimeter Institute, of which I am a member, that there is momentous news being announced today at a press conference. The rumor is that gravitational waves have been detected for the first time.

We gather in the Perimeter Institute's bistro in front of a giant screen. Dozens of theoretical physicists are chattering excitedly, some of them fiddling with their laptops, as we wait for the screen to light up. Sitting in anticipation of the press conference, I gaze at the bleak winter landscape outside the floor-to-ceiling windows in Waterloo, Ontario, Canada. A small gaggle of Canada geese meanders by, looking dignified despite the cold.

The Black Hole Bistro feeds the more than 100 theoretical physicists and administrators who occupy the offices of the large, modern building, with its dramatic angles of glass and black-and-gray walls, overlooking small, scenic Silver Lake. The intense brain activity at the Institute by theorists struggling to understand the mysteries of the universe can, at times, be palpable. Some members of the Institute will have their careers significantly enhanced if today's press conference does announce that gravitational waves have been detected.

Suddenly, a picture flashes on the screen of a panel of senior members of the LIGO team and a picture of the auditorium in Washington, DC, where the press conference is being held. We can see television cameras and information technology operators moving around wearing headphones.

After an introduction by France Córdova, the head of the National Science Foundation, the major funding agency, a distinguished, gray-haired gentleman steps up to the microphone. He is David Reitze, LIGO Laboratory Executive Director.

"We have detected gravitational waves," he announces simply. "We did it!"

The physicists in the bistro all applaud the announcement, along with the cheering audience on the screen. A current of excitement ripples through the bistro as we absorb the significance of this statement.

HOW DID THEY DO IT?

Different presenters at the press conference give us a picture of the gravitational wave detection. The actual detection of GW150914 (which stands for "Gravitational Wave 2015 September 14) occurred through a series of events. Binary inspiral search algorithms running at the two observatory sites—in Hanford, Washington, and Livingston, Louisiana—were triggered by the incoming data. The computers sent that metadata to computers at Caltech in Pasadena, California, which identified the event. The time–frequency plots of the distinctive "chirp" in the data were uploaded automatically at about 2:55 a.m. Pacific time, and were first noticed by postdoctoral fellows in Germany, who were manning their computer screens at that time of day. The "chirp" consists of a rise in the frequencies of the gravitational waves from about 30 Hertz to 400 Hertz. The chirp is a measure of the strength of the amplitude of the gravitational waves emanating from the coalescing black holes.

The merging of two black holes weighing in at 36 and 29 solar masses began as they approached one another at a distance of only about 350 kilometers. At this point they were whizzing around one another at about half the speed of light. The actual collision or merging occurred within two tenths of a second, and it produced a black hole of about 62 solar masses. The missing 3 solar masses are a result of the release of 3 solar masses of energy in the form of gravitational waves. Recall from special relativity that energy $E = mc^2$, Einstein's famous formula. For a very brief time, this enormous energy output is greater than the energy output of all the stars in the heavens! Such a cataclysmic release of energy produced a quivering of spacetime, and this quivering traveled out in all directions in the universe. It took 1.4 billion light years for the ripples to reach Earth, where they made the space in between the test mass mirrors in the LIGO apparatus tremble ever so slightly, producing the tiny strain $\Delta L/L$ equal to 10^{-21}. This tiny movement revealed dramatically the detection of gravitational waves.

Relativists in the past, using Einstein's theory of gravitation applied to compact binary systems, calculated that when gravitational waves are produced from such a binary system, the frequency of the waves would be twice the orbital frequency of the swirling compact objects, which could be black holes or neutron stars. They then estimated that just before the merging of the objects, the observed wave frequency would be about 200 or 300 Hertz, from which it could be inferred that the compact objects were orbiting each other about 100 times per second. From the strength of the observed amplitude of the waveform and the masses of the black holes, the distance from Earth to the black hole merging event could be determined to be about 1.4 billion light years. From the masses of the black holes, we can also determine the speed of the orbiting black holes. The smaller the masses, the slower the orbital speed, and the higher the frequency of the gravitational wave.

How are the masses and spins of the black holes determined? How much evidence for gravitational waves can be obtained from the actual observed raw waveform data? The answer to the first question relies on producing computed waveforms consistent with the predictions of general relativity. A large library of these computed waveforms, called *templates*, has been produced over several years (Figure 7.1). Each template corresponds to some assumed masses and spins of the inspiraling black holes. These template waveforms are compared to the observed waveforms, taking care to eliminate sensitivity to noise such as seismic events, instrument calibration signals, and any other conceivable form of background signal.

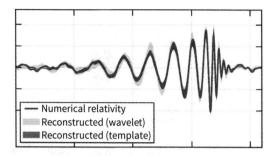

Figure 7.1. Gravitational wave template obtained from numerical relativity. The template fits the first event, GW150914; the data can be seen as a shadow underneath the bold line of the template. Credit: LIGO

The template waveforms have only been calculated assuming general relativity is correct. To check whether an alternative gravity theory predicts deviations from general relativity requires a whole new library of waveform templates for each theory.

The answer to the second question is that a fair amount of information can be obtained from the data after software has been used to clean the data of noise. The details of the technology are complicated; but, simply put, the cleansing of the data pulls out a signal-to-noise ratio. If this ratio is sufficiently large, then it can be claimed that a real gravitational wave signal has been detected. Such an analysis of the first event, GW150914, produced a large enough signal compared to the background noise that the advanced LIGO scientists could claim with certainty that gravitational waves had been detected.

The library of templates deduced from numerical calculations of Einstein's field equations contains several hundred thousand templates, which describe every conceivable combination of mass and spin of the merging black holes that could produce the observed waveforms. For GW150914, the first event, the best matching template picked out the two masses of the black holes at 36 and 29 solar masses.

Determining the spins of the black holes is a greater challenge. These are indicated by a special spin parameter that can determine the alignment of the spins to the binary orbital angular momentum plane, or the

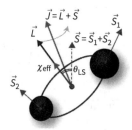

Figure 7.2. The merging of binary black holes with spins S_1 and S_2, with L and J the orbital angular momentum and the total angular momentum, respectively. The two black holes are spinning in the same direction as their orbit (i.e., S and L are parallel). The χ_{eff} (chi effective) is the measured spin quantity using the sum of the black hole spins, S, and the orbital angular momentum L. We expect that dynamically formed binaries would have their spins aligned at random, so we do not expect to see many black hole binaries with S and L parallel. Credit: "Black Hole Spins" by Carl Rodriguez on his website, *BHDynamics*

magnitudes of the spins of the two binary compact objects—in this case, black holes (Figure 7.2). A spin alignment vector is combined with the orbital angular momentum vector to produce an effective measurement of the spins of the black holes. The measurements showed there were two possibilities for the spins of the black holes. Because the effective measurements of the spin parameter were clustered around zero or were negative for the gravitational wave events detected, this meant either that the spins of the black holes were small (i.e., they were not rotating much at all) or that the spins were misaligned with respect to the orbital angular momentum vector. Currently, with the available gravitational wave events, we are not able to decide which is the correct interpretation.

GETTING INTO THE HABIT OF DETECTING GRAVITATIONAL WAVES

The second gravitational wave event detected by advanced LIGO was at first called LVT151012, where LVT stands for LIGO/Virgo transient. The probability for this being a gravitational wave event was not as strong as with the first event, GW150914. However, it was eventually upgraded to official gravitational wave catalog status, and was renamed GW151012. Of the first six gravitational wave detections by the advanced LIGO/Virgo detection network up through the end of 2017, the first event produced the strongest signal, confirming the existence of gravitational waves, caused by the merging of two black holes. The other five events had weaker signals, with a match to template waveforms that could not be declared nonrandom quite so confidently. Nevertheless, it is believed they are true gravitational wave events.

The third event, discovered at Livingston and Hanford on Boxing Day, December 26, 2015, had a stronger probability of being nonrandom than the second event, LVT, despite a weaker signal. Indeed, the third event, reported by LIGO on June 15, 2016, called GW151226, was a dramatically important announcement because it verified that the first event was not a fluke. The analysis of the waveform data by the LIGO/Virgo collaboration revealed that the Boxing Day event was also the merging of two black holes. The masses of the two black holes in this system were 14 times the mass of the Sun and eight times the mass of the Sun. Their spins were not as well determined in the analysis as in the case of the first event. But, the fact that two gravitational wave events detected with sufficient confidence had been recorded by the experiments validated the expectation that there would be many more to come in future experimental runs of the detectors starting in fall 2016.

The smaller black hole masses in the third binary black hole system GW151226 are more in agreement with the measured masses obtained from observations of X-ray black hole binaries in our galaxy. The distance of this second source was approximately the same as with the first system—namely, about 1.4 billion light years from Earth.

The three additional events detected by LIGO/Virgo in 2016 and 2017 provided more confirmation of the detection of gravitational waves, even though the signal strengths of these events were not as strong as the first event. The seventh event, detected by the combined LIGO/Virgo collaboration in 2017 was the result of the collision of two neutron stars. This exciting discovery is covered in Chapter 8. At this writing, there are more than 12 detections of gravitational waves from the first and second runs

(01 and 02), 11 from black hole binaries, and one neutron star merger. The LIGO/Virgo project has been upgraded and commenced operating again in 2019. We expect, as a result of the increased sensitivity of the experimental setup, that many more gravitational wave detection events will be discovered.

WHAT ARE GRAVITATIONAL WAVES?

As we have learned, gravitational waves (not "gravity waves," which are waves in fluids, such as the oceans and atmosphere, and are the result of gravity) were first predicted by Einstein's theory of general relativity in 1916. They are caused by the acceleration of a massive object within spacetime, which stretches and compresses spacetime in a wave-like motion. Because they are coming from so far away, the gravitational waves detected by advanced LIGO/Virgo involve incredibly small distortions of spacetime. Indeed, in Einstein's papers in 1916 and later in 1918, he asserted that gravitational waves should exist, but that they would be impossible to detect. The fact that LIGO/Virgo is now detecting these waves is astonishing. It is particularly amazing that the cause of the minuscule distortion of spacetime is the collision of two black holes or neutron stars in binary systems so very far away from Earth. Recall that a distance of 1.4 billion light years means that the collision took place *1.4 billion years ago*, when on Earth the only life forms were single-cell or primitive multicellular organisms.

In general, any two merging black holes detected by LIGO will combine in a violent and catastrophic collision. Such black holes were born billions of years ago as collapsing stars. They were originally well separated in their orbits around one another. The loss of gravitational wave energy from the binary system slowly reduced the distance between the black holes and they gradually approached one another. As they approached one another, they sped up in their orbital motion as a result of their mutual gravitational attraction. When they got very close, they were spinning around each other at more than half the speed of light.

During this time of inspiraling motion, the frequency of the gravitational waves increases until the dramatic moment when the black holes crash into one another. The initially merged system is highly excited, although the excitation energy rapidly dissipates, being converted into gravitational waves, and resembles a strongly damped bell. The oscillations of the ringing bell form what is called the *ring-down phase*, which contains the characteristic frequencies of the quasi-normal modes. The system finally forms a single, quiescent black hole. This all happens within a very short time—fractions of a second—and now the frequency of the gravitational waves enters the audible range. The pitch of the sound increases, and it can be heard as an audible rumbling. As the two black holes merge into one, the pitch produces the characteristic "chirp" (Figure 7.3).

As Mike Landry, head of the Hanford LIGO laboratory, explained later during our visit to the site, we do not actually hear the coalescing of the black holes. It is only when the data are analyzed digitally that we can convert the rumbling and the final chirp to audible sounds. The conversion of the signal to an audible sound can be compared to how an electric guitar makes music. The strumming of the guitar strings is passed to magnetic pickups, which detect changes in the field of a permanent magnet caused by motion of the nearby conducting guitar strings. The amplifier then produces music in the audible Hertz range of frequency.

The resulting black hole is roughly the size of the sum of the masses of the two black holes less the amount of mass/energy that was radiated away by the gravitational waves.

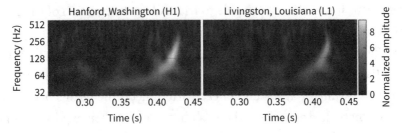

Figure 7.3. Time–frequency plot that shows the signal of the binary black hole merger increasing with time and creating the characteristic "chirp." Credit: LIGO

THE INNER WORKINGS OF THE LIGO DETECTOR

My wife Patricia and I timed our visit to the LIGO site in Hanford in March 2017 to be present during the maintenance period on a Tuesday morning, when the engineers perform tests and updates on various components of the detection system. Such LIGO maintenance takes four to six hours. The liquid nitrogen that cools the traps to freeze out hydrocarbons and effectively "pump" out gas is refilled. Contractors fix things, mats are replaced, and debris is removed from the enclosures surrounding the 4-kilometer beam tubes.

"Mice are a big problem in the beam tube covers, and their impact on the external surfaces of the beam tubes has to be cleaned," Mike said, as he led us carefully over colorful cables on the floor. "Even mouse urine is a problem in the steel tubes." The maintenance workers frequently remove mice, and once they even had to remove a snake, Mike explained.

Mike gave us a guided tour of the major installations within the detector (Figure 7.4). He took us to the vertex of the two right-angle 4-kilometer arms, which we viewed standing above them on an iron platform. Part of the equipment we viewed will be shipped to the upcoming LIGO project in India, due to start up in 2025.

"The Italian gravitational wave detection project called Virgo is coming online in July 2017, and we will run in conjunction with them until August," Mike said, adding there will also be a gravitational waves project called KAGRA in Japan later. The more detectors located around the world, the better the resolution and sensitivity in the detection of gravitational waves. In addition to increased resolution, having many detector sites means it will also be possible to triangulate the position of the gravitational wave source out in space more accurately.

Mike showed us the external housing for the vacuum chamber that splits an initial laser beam into two beams, which then travel the length of the two arms. A mirror at the end of each arm reflects the beams back to the vertex, where they are reflected into a photodiode. When exposed to light, the photodiode produces an electrical potential, which converts any signal into a recognizable fringe pattern. The glass mirrors, or test masses, at the

Figure 7.4. Tour of LIGO with Mike Landry. Credit: Patricia Moffat

ends of the arms are large and thick; they weigh 40 kilograms each, are 34 centimeters across, and 20 centimeters thick. These test mass mirrors are suspended by thin fibers of amorphous fused silica that vibrate like a violin string. The laser beams cause the optic mirrors to vibrate at 15 to 120 Hertz. The input test masses are a few meters beyond the beam splitter (Figure 7.5). This means the paired input test mass and the end test mass of a given arm form a Fabry-Pérot optical storage cavity of 4 kilometers in length.[1]

Rick Savage, the calibration specialist at LIGO Hanford, explained to us that the glass fibers suspending the mirrors can explode if you touch

---

1. A Fabry-Pérot interferometer is an optical cavity made from two parallel reflecting surfaces composed of thin mirrors. Optical waves can pass through the optical cavity only when they are in resonance with it. Charles Fabry and Alfred Pérot developed the instrument in 1899.

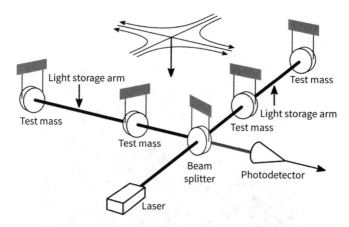

Figure 7.5. LIGO interferometer detector apparatus. Credit: LIGO

them, simply because of the oils on your finger. This would be a costly mistake, because it takes six weeks to fix these fibers.

To detect a gravitational wave, the purest, most coherent laser light is needed. A pure, narrow infrared laser is used, and the light is passed through a crystal before it enters the machine. The crystal phase modulates the laser light. The whole system is highly automated, and when the system is "locked," it is resonating—or working—whereas when it is "unlocked," it is not resonating and is not ready to detect a gravitational wave.

At the vertex of the arms, the interference patterns of the two returning laser beams are digitally analyzed by computers. If one of the arms has been displaced a tiny amount, by a gravitational wave or a faraway earthquake or other external causes, then there will be a very small but detectable change of phase of the wave pattern in the laser light. This displacement in the tube inside the interferometer arm is extremely small, about 10^{-21} times 4 kilometers, which is about 4×10^{-16} centimeters. This is, in fact, too small to picture easily; recall, it is about one one-thousandth the diameter of a proton.

In the absence of gravitational waves, the laser beams traveling in the two arms arrive at the photo detector nearly out of phase, thereby producing no signal. A gravitational wave propagating perpendicularly to the plane of the detector spoils this perfect destructive interference of the laser beams and produces a signal. What the detector at the vertex sees is a "fringe pattern," caused by the interference of the laser light waves. The interference pattern can be positive or negative, depending upon whether the crests of the waves hit one another or whether a crest hits a wave trough.

During a detection, the machine does not capture only one gravitational wave oscillation. Like a tossed pebble creating concentric circles of waves in a pond, gravitational waves, too, come in groups.

"About 50 gravitational waves come in when a signal is detected," Vern Sandberg, the engineering and electromagnetic specialist, told us later. (Vern's earlier career in physics was as a theoretical physicist specializing in relativity theory.)

Thus, gravitational waves move in cycles. During the first half cycle, the waves lengthen one arm and shorten the other, whereas during the second half cycle, these changes are reversed. The variations of these lengths change the phase difference between the laser beams and produce an optical power signal that reaches the photo detector at the vertex of the L as a fringe pattern.

As Mike Landry explained, "The lighter the system, the longer it is within our sensitive band. GW150914 was about a quarter second in the detector. GW170817 lasted 100 seconds in the detectors."

When a gravitational wave passes through the LIGO interferometry system, it distorts the geometry of space, changing the distance between the mirrors a tiny bit, and this movement, called the *strain* of spacetime, changes the fringe pattern detected at the vertex of the L. This minuscule change signals that a gravitational wave has been detected from a distant source. The analysis of the fringe data converts it into a wiggly waveform. The sizes of the wiggles increase in amplitude just before the two black holes coalesce, and then decrease and finally end as tiny wiggles during the ring-down phase when the two black holes form a single black hole that is reverberating like a ringing bell. The final dramatic increase in the wave motion and the ring down of the waveform happens in milliseconds. Recall that at the merging of the black holes, the frequency of the gravitational waves has reached audible levels in the region between 30 Hertz and 150 Hertz. The chirp is heard as a rapid increase in the pitch of sound.

Remarkably, this sound chirp is also what is heard during large earthquakes. There is a sound recording of the 2011 earthquake in Christchurch, New Zealand, which measured 6.3 on the Richter scale, where the same chirp sound can be heard. However, earthquakes occur at frequencies around a tenth of a Hertz, and not in the range of 100 Hertz, as in the LIGO gravitational wave detection, so gravitational waves can be discriminated from earthquakes.

LISTENING THROUGH THE NOISE

During our conversation with Rick Savage, the senior experimental physicist, he explained in detail how technicians manage to isolate the interferometer from any external noise or thermal distortions that could mimic a gravitational wave signal.

I asked him, "In view of the minuscule displacement of one of the arms of the interferometer by the amazing one thousandth of the diameter of a proton, would this not introduce significant quantum mechanical physics into the analysis of the data?"

"Good question," Rick said. "As it turns out, it's the averaging of all the tiny cells of deformation producing a macroscopic nonquantum mechanical signal that we actually observe. In this way, we avoid any difficulties with minuscule quantum mechanical effects despite the tiny displacement of the detector arm." The physics behind this averaging thereby avoids the necessity to involve poorly understood quantum mechanical effects at such tiny distances.

Then I said, "Aha, so this avoids having to deal with the standard model of particle physics and the fact that the proton is made up of three quarks."

The detection of the first gravitational wave first occurred at the Livingston, Louisiana, site and about seven milliseconds later at the Hanford LIGO site in Washington. This difference is because the gravitational wave moved from the Livingston site to the Hanford site at the speed of light, which took time to travel. The coincidence of these two detections at very similar waveforms proved this was not an Earth-based seismic event, but was coming from outer space.

The sensitivity of the LIGO detector is remarkable. However, the biggest problem facing the LIGO collaboration is removing noise, the primary sources of which are seismic waves, thermal motion within the instruments themselves, storms, lightning, and movement of vehicles and even people at the site. Some of these disturbances one would not even imagine. Mike Landry explained to us that a significant source of noise is during storms, when strong waves crash against the submerged continental shelf many miles away in western Washington.

DETECTION COINCIDENCES

In 2015, the LIGO detector underwent a second upgrade to improve its sensitivity by a factor of 3 to 5 for waves in the 100- to 300-Hertz frequency range, and by about a factor of greater than 10 for frequencies less than 60 Hertz. These increases in detector sensitivity to more distant sources of gravitational waves were the turning point in the discovery of gravitational waves from a source that was more than a billion light years away. Within the first three days of the LIGO detector being turned on in a stable configuration, on September 14, 2015, a signal was detected that was so strong that the chirp signal within the detector output could be seen by the eye, "once bandpass and notch filters for 60-Hertz mains were applied to the data," Mike Landry added. The important part of the signal corresponding to the merging of the two black holes lasted only about two-tenths of a second, and it was detected in both detectors, first at Livingston and then at Hanford. At the time, workers at both sites thought that perhaps this was a fake signal, or a blind injection signal test inserted by management, because the machine had so recently been turned on for the first time.

The second and third events happened on October 12, 2015, and on Boxing Day, December, 26, 2015. The machine had not been operating on Christmas day, Mike Landry said, as we stood beside the beam splitter wearing our visitors' hairnets and slippers. Then, when it started up again, a signal was detected early in the morning of December 26th. *A Christmas present!* Mike had immediately thought, after Vern Sandberg contacted him in Victoria, British Columbia, during the Christmas holidays. There had been the earlier event on October 12, 2015, but the consensus was that it was not a strong-enough signal, at 1.7 sigma, to take seriously. It could have been just a random fluctuation in the noise that happened to look like a gravitational wave.

Again, like the earlier detection, the fourth event, which occurred not long before our visit to LIGO, on January 4, 2017, happened shortly after the machine had been turned on after a period of rest; in this case, bad weather had caused it to be shut down. It almost sounds like the machine is eager to prove its worth after a period of rest.

INTERPRETING THE DATA: WHERE ARE THE WAVES COMING FROM?

How do we know that the gravitational wave signal for the GW150914 event was coming from two distant black holes of 36 and 29 solar masses, rather than, say, from two neutron stars? Recall that the conclusion is based on a numerical solution of the field equations of Einstein's general relativity that simulates just such an event, using highly technical numerical methods developed over more than 30 years. These numerical calculations produce as an end result the fringe and waveform patterns that can be compared to those observed by the Livingston and Hanford advanced LIGO systems. The numerical models produced more than 200,000 waveform templates for different combinations of mass and spin that could be compared with the observational data.

The template that fits the data best tells us the masses of the two components of the black hole binary system and the magnitudes of the spins of the two black holes. It also tells us the mass and spin and angular momentum of the final black hole created by the merging of the two black holes. The parameter estimation teams go further, interpolating between template values. Ultimately, the best-fit parameters are obtained through numerical relativity.

We know from general relativity and nuclear physics that a neutron star cannot possess a mass greater than about three times the mass of the Sun. This mass limit, of course, is way below the 30-some solar-mass compact objects that produced the GW150914 gravitational wave signal.

However, a possible conflict arises between the predicted masses and spins of the two black holes, and observational data concerning the mass, spin, and angular momentum of the final black hole. The theory and observations are at odds. The only direct observational data we have for the masses of black holes come from the electromagnetic radiation processes associated with X-ray binary black hole systems. These observational data place a bound on the black hole masses of a large number of X-ray binary systems observed in our galaxy. This mass bound says that a black hole cannot have a mass bigger than about 10 times the mass of the Sun. This is three or more times smaller than is needed to fit the predictions of general relativity to the new LIGO data.

We also have to understand the formation processes of stellar-mass black holes if they are to have masses bigger than 10 times the mass of the Sun. The development of models of black hole binary system formation has a long history going back more than 30 or 40 years, and the possible formation channels are perhaps in conflict with the 30 solar-mass black holes that are needed by general relativity to complete the description of the GW150914 gravitational wave detection event. However, if such larger mass black holes exist, then it will signal that they can be found billions of light years from Earth, rather than in our galaxy. Another possibility is that the massive stellar black holes are primordial black holes formed in the early universe.

Another possible problem is the observed spin and angular momentum of the two merging black holes. The measured effective spin parameter is too low to agree with what one expects for the angular momentum or spin of the two original black holes, and the final merged black hole. After billions of years, the inspiraling black holes are expected to have spins that are aligned, not significantly antialigned, as the data suggest, although the other possible explanation is that the black holes are rotating slowly.

There are four possibilities for the formation of the merging black holes. One is that the black holes formed a binary system when two stars that were already in a binary system collapsed (most likely at different times). This binary system slowly inspirals over billions of years and eventually the two black holes collide and merge into one larger black hole. This merging black hole scenario prefers that the spins of the two black holes be aligned with the plane of the orbital angular momentum.

A second evolutionary model has a black hole capturing another black hole and forming a black hole binary system. In this case, the merging of the black holes would occur faster than in the first evolutionary model, and the merging black holes could have different spins and be misaligned with the orbital angular momentum of the binary system.

A third possibility for the formation of binary black holes is that they have formed by a supernova explosion that produced two black holes. There exist models describing this possible birth of a binary black hole.

A fourth possibility for the formation of binary black holes is that they are formed in a dense environment, such as in a globular cluster or at the center of a galaxy. In this case, the spins can be aligned or misaligned.

The magnitudes of the spins measured in X-ray binary systems in our galaxy are substantial, and the spins of the black holes are found to be preferentially aligned with the orbital angular momentum. This would tend to cast doubt on the numerically measured value of the effective spin of the LIGO events, which could be significantly smaller in magnitude than those in our galaxy.

Why would the spin properties of the binary black holes be significantly different in galaxies more than a billion light years from ours? This is an exciting, important conundrum that has yet to be solved. If 15 or 20 more events are detected by the LIGO/Virgo collaboration that all have misaligned spins and a small or zero value of the effective spin parameter, then this raises an issue about the formation mechanism of binary black holes, or possibly a problem with the analysis of the LIGO data.

One possible explanation for the large masses required by the data and general relativity, if you recall, is that the black holes could be so-called *primordial black holes*, created during the very early universe right after the Big Bang. However, existing data from WMAP and Planck satellite experiments appear to exclude the existence of primordial black holes with a mass between 30 solar masses and 100 solar masses. Usually, primordial black holes are considered to be of a tiny mass, about 10^{-13} times the mass of the Sun or less; these are known as *mini black holes*.

To resolve the tension between the astronomical observations—no black holes more massive than about 10 times the mass of the Sun—and the LIGO/Virgo data, which require black holes of larger masses, some theorists have begun speculating about the existence of much more massive primordial black holes. These would be formed by the collapse of matter density fluctuations in the very early universe, shortly after the Big Bang, whereas the expected binary black holes are formed by the collapse of massive stars or by supernovae explosions in the current universe. More accurate data are required to constrain the sizes of massive primordial black holes.

On August, 1, 2017, the Virgo gravitational wave observatory near Pisa, Italy, began operations, and is now working in tandem with the Hanford and Livingston observatories to detect gravitational waves. The arms of the Virgo interferometer are 3 kilometers in length, which is 1 kilometer less than their U.S. counterparts. The sensitivity of the Virgo detector is therefore less than at the U.S. sites. Nonetheless, the LIGO/Virgo collaboration can use the Virgo interferometer to help triangulate the sources of detected gravitational waves, and help to improve the reduction of noise in the gravitational wave detections and their analyses. The detection of the sources of the gravitational waves would significantly increase our understanding of the kinds of scenarios that lead to the merging of two black holes.

NUMERICAL RELATIVITY CALCULATIONS (FOR MATHEMATICALLY INCLINED READERS)

To obtain a deep understanding of gravitational waves, we must be able to solve the field equations of general relativity and any possible alternative theory of gravitation. Unfortunately, general, analytical solutions of the field equations are difficult to obtain because, in general relativity, the field equations constitute 10 nonlinear partial differential equations. In contrast to this situation, Maxwell's equations constitute linear partial differential equations, which are much easier to solve to obtain descriptions of electromagnetic phenomena.

However, there are the exceptions to Einstein's field equations when exact solutions can be obtained, such as the Schwarzschild solution yielding black holes, the Kerr solution of a spinning black hole, and the Kerr-Newman solution for electrically charged black holes. These three are exact solutions that can be derived because certain symmetries of spacetime are assumed. For example, if we assume that the astrophysical body we are trying to find a solution for is exactly spherically symmetric, then from the vacuum field equations in which matter is absent, the field equations can be solved, as was first demonstrated by Karl Schwarzschild in 1916. For this solution, it also has to be assumed that the gravitational field vanishes at infinite spatial distance where spacetime is described by a flat Minkowski spacetime metric.

In the case of the Kerr solution, the assumption consists of imposing axial symmetry about the z axis in three dimensions (x, y, z), describing a body that has angular momentum and spin about that axis. It is also assumed that for the Schwarzschild and Kerr solutions, the solution is "stationary" or independent of time.

Another solution of Einstein's field equations was obtained by Alexander Friedmann, when he assumed that the universe is both homogeneous and isotropic, as was originally assumed by Einstein in his first paper on cosmology, published in 1917. In the language of differential equations, the homogeneity of spacetime means that the matter density distribution is spatially uniform everywhere, and the density is only dependent upon time as the universe expands in time from the Big Bang. The isotropy assumption, on the other hand, holds that the universe looks the same for observers viewing it in any direction. We also must take into account the pressure of matter, and—with density—the pressure also only depends upon time from the Big Bang onward. Because of these simplifying assumptions, Friedmann was able to solve Einstein's field equations analytically and exactly, and the resulting equations are called the *Friedmann equations of cosmology*.

There is another exact solution of the field equations found by George Lemaître and independently by Richard Tolman, which was further developed by Hermann Bondi, for which the mathematical differential equations can be solved for non-uniform matter without pressure being exerted between the matter particles.[2] The solution is also spherically symmetric. This solution is called an *inhomogeneous cosmology solution* and has been proved useful as a model of the large-scale structure of the universe.

For example, in 1992 and 1995, I published—in collaboration with a graduate student, Darius Tatarski —articles on what I called a *void cosmology*.[3] Large voids in the universe have been observed, surrounded by filaments of matter. These voids can be as large as almost a billion parsecs in size, where 1 parsec is equal to 3.26 light years. In void cosmology, we picture that our Earth in the Milky Way galaxy is near the center of one of these very large voids with little matter in it. That is, we are near the center of a large, spherically symmetric bubble. In this model, the acceleration of the expansion of the universe discovered from supernovae observations in 1998 is an illusion. Light emitted from supernovae and passing through the void differs from light passing through dense matter outside our bubble. Thus, to us, the supernovae appear to be farther away

2. G. Lemaitre, "L'univers en expansion," *Ann. Soc. Sci. Bruxelles I*, **A53**, 51–85 (1933); R.C. Tolman, "Effect of Inhomogeneity on Cosmological Models," *Proceedings of the National Academy of Science U.S.*, **20**, 169–176 (1934); H. Bondi, "Spherically Symmetrical Models in General Relativity," *Monthly Notices of the Royal Astronomical Society*, **107**, 410–425 (1947).

3. J.W. Moffat and D.C. Tatarski, "Redshift and Structure Formation in a Spatially Flat Inhomogeneous Universe," *Physical Review*, **D45**, 3512 (1992); J.W. Moffat and D.C. Tatarski, "Cosmological Observations in a Local Void," *Astrophysical Journal*, **453**, 17–24 (1995).

than they actually are. The fact that void cosmology assumes we are near the center of the spherically symmetric bubble contradicts what is called the *Copernican cosmological principle*, which says there is no special place or time anywhere in the universe; that is, we do not occupy a special place in space or time. It is difficult to prove the Copernican cosmological principle observationally, for we would have to make observations far away from our galaxy, which is clearly not easy to do.

When we attempt to solve the field equations of general relativity for two coalescing binary black holes, there is insufficient symmetry to allow us to obtain an exact, analytic solution of the field equations. The only way to solve the equations is to turn to numerical methods, using large-scale digital programming code. Attempts to solve the equations numerically were first made during the early 1960s, but these attempts were crude because there wasn't sufficient computer power available at the time. Since then, supercomputers have developed to a remarkable degree with regard to speed and memory capacity, and we are now in a position to find numerical solutions of the gravitational equations of general relativity for exotic physical phenomena such as the merging of two black holes.

Despite the significant increase in computing power, physicists attempting to solve Einstein's field equations have been faced with formidable problems. The computer code and algorithms used to solve the field equations have often failed as a result of severe instability problems. It is only during the past few years that these problems have been overcome, and physicists such as South African physicist Francis Pretorius, now at Princeton University, succeeded in devising computer code that avoided an instability. No longer did the simulation software crash when two black holes were in the final stage of coalescence. This opened the door for collaborations of physicists doing numerical relativity with those analyzing the actual fringe patterns observed by the advanced LIGO collaboration. The successful comparison of the calculated fringe–waveforms from general relativity and those observed at LIGO provided direct verification of the existence of gravitational waves.

The remarkable success of the numerical relativity research over a period of two or three decades has opened up an exciting new vista for understanding the physics of strong gravitational fields. For the merging of the two black holes, and their emission of gravitational waves, the numerical relativity solutions of Einstein's equations produced oscillating waveforms, which can be used as templates for comparison to the observed statistically analyzed waveforms describing the merger. Recall that this has generated a library of about a quarter of a million template solutions of Einstein's equations. By analyzing these templates, the best template fit to the waveform data was used to determine the masses and spins of the two merging black holes and of the final quiescent black hole.

Extensive numerical relativity calculations will also help interpret the upcoming LISA observations. LISA is, effectively, a much larger LIGO and Virgo experiment in space. These numerical relativity calculations will allow the physics community to plumb such mysteries of the universe as the Big Bang. The LISA experiment will permit the detection of gravitational waves at significantly lower frequencies of 10^{-10} to 10^{-5} Hertz. At these frequencies, we can detect gravitational waves being emitted by supermassive black holes at the centers of galaxies and, ultimately, the gravitational waves emitted at the Big Bang.

MORE DETECTIONS OF GRAVITATIONAL WAVES AT LIGO

The official announcement of the fourth detection of gravitational waves was on June 1, 2017 in *Physical Review Letters*.[4] It was called GW170104 because it was detected on January 4, 2017. This was the fourth event, if one counts LVT151012, which was initially dismissed by the LIGO collaboration. The article made for interesting reading, considering that we had been present at an editorial meeting for this article in March when we visited the Hanford observatory.

This gravitational wave signal was detected with high statistical significance. The data were represented by a time–frequency waveform.

4. B.P. Abbott et al., "GW170104: Observation of 50-Solar-Mass Binary Black Hole Coalescence at Redshift 0.2," *Physical Review Letters*, **118**, 221101 (2017).

The waveform was clear and reasonably noise free, as was the waveform of the first detection, GW150914, which had an even higher statistical significance. When converted digitally, the signal was visible as the characteristic chirp of a binary coalescence of two compact objects identified as black holes. An analysis of the data for GW170104 demonstrates that the signal arrived at Hanford about seven milliseconds before arriving at Livingston, and originated from the coalescence of two stellar-mass black holes at a luminosity distance from Earth of about 3 billion light years. The luminosity distances for the two previously statistically significant gravitational wave detections were about 1.5 billion light years. The inferred masses of the two black holes are about 31 and 19 solar masses—larger than expected, just as in the first two gravitational wave detections. The source of the colliding black hole system had a total mass of about 50 solar masses.

The calculated merger rate of black holes from the recent detections—that is, how often do black holes coalesce?—is consistent with previous calculations for the other detections. These probability calculations are based on theoretical models for the evolution of binary compact systems and their coalescence. The merger rate of black holes in the universe is, in a sense, an educated guess, estimated by the first detection of several events every billion years. It can be explained by binary black holes forming through isolated binary evolution or, alternatively, through dynamic interactions in dense stellar clusters. We do not yet understand the dynamic interaction of the binary system with other astrophysical objects in dense stellar clusters. Such interactions would affect the coalescence rate, compared to binaries in a nondense environment. The problem is that the large masses of these stellar black holes detected by LIGO have to be explained through evolutionary models and, recall, this is currently a topic of some controversy.

The black holes in any inspiraling binary system are formed through the collapse of the progenitor stars. To leave a heavy black hole as a remnant, the progenitor stars must avoid significant mass loss. If the binary system is subjected to what are called *stellar winds*, then the constituent black holes lose mass—a situation that is not compatible with the larger black hole masses of the detected binaries. Also, if the progenitor stars have too much heavy metal in them, then there would be too much mass loss, and they would not produce the heavy masses being inferred from the data.

As I have discussed, another possible stellar black hole formation channel would be the existence of primordial black hole binaries in which the black holes are much larger than astrophysicists have previously thought. The masses of the detected mergers so far lie in a range for which larger primordial black holes could exist, and it is speculated that these primordial black holes could contribute significantly to the existence of dark matter in the universe. However, satellite data and lensing data have so far produced bounds on the masses of these dark matter primordial black holes, making their existence doubtful.

A further issue is the measurement of the black hole spins, which is still a problem after the binary black hole detections in the second run of the LIGO/Virgo detectors. The model simulations of inspiraling black holes over a period of 9 billion years would, as a result of mass losses via gravitational radiation during the inspiraling phase and tidal interactions, prefer that the black hole spins are approximately vertically aligned, like two dancing skaters rotating around one another. The observed merging black hole spin misalignments, however, are more like dancing skaters who are twirling around one another with one or both skaters leaning back toward the ice. The black hole spins play a role in the orbital evolution of the binary system and are not easily determined. The expected spins of the two black holes should be aligned with the orbital angular momentum. However, the detected gravitational wave signals show they are misaligned, or, alternatively, the black hole spins are small.

Let us step back for a moment and consider that our planet Earth was formed about 4.5 billion years ago, which is more than a third of the age of the universe. The event that produced the fourth gravitational wave occurred 3 billion years ago, when the Earth was 1.5 billion years old. Let us think about what was happening on our planet at that time. The Proterozoic Eon is a geologic time period of about 2.5 billion to 500 million years ago, or the latter half of the Precambrian Era. During this time, simple life forms such as bacteria, blue-green algae, and the first simple oxygen-dependent animal life began evolving into more complex living organisms. The chirp of the gravitational waves detected in January 2017 originated during that time. Even if there had been someone present on Earth then to hear the gravitational wave breaking, they would not have been able to detect it for another 1.5 billion years.

PROPERTIES OF THE LIGO DATA

As we have learned, there are three phases exhibited by the observational data obtained by the LIGO observatories: (1) the inspiraling phase of the two compact objects (black holes or neutron stars) or supermassive black holes, (2) the merging of the two objects, and (3) the ring-down phase, when the two objects form the final black hole or neutron star. The observational data obtained by the LIGO observatories span 4096 seconds for the first event to complete the three phases of merging. Less than one second of signal is at a frequency with sufficient amplitude to be distinguished from the noise. For the typical time record, much of these data are not useful for interpreting the merging of the black holes for the other detected events. Except for a few seconds, the inspiraling event data are at a frequency of 7 Hertz, at which the noise in the data makes it currently impossible to interpret. As the black holes come together during the inspiraling phase, the frequency of the waveforms is increasing; when it exceeds 30 Hertz, the data can be usefully interpreted. The inspiraling part of the waveform obtained from the strain measurements takes the approximate form of a sine wave, but only tens of cycles are useful for analysis. The merging of the black holes takes place in about two tenths of a second, and these data are at a frequency of 100 Hertz or more, and are a significant part of the analysis of the data. The ring-down phase is very short, and these data require many gravitational wave detections to be useful in interpreting the data. As the LIGO observatories' data sensitivity increases with planned upgrades, more useful data for the inspiraling phase and the ring-down phase should become available.

General relativity is used to interpret the data. The numerical calculations, which consume considerable amount of computer time, have been performed over the past several years. Of the quarter of a million templates, those that fit the strain wave indicate the likely masses and spins of the merging black holes. Currently, the best theoretically obtained waveform shows the data are consistent with general relativity. It should be cautioned that consistency does not constitute proof that general relativity is correct.

Alternative gravity theories have free parameters, whereas general relativity has no free parameters. (My alternative gravity theory, MOG, has only one free parameter for black hole mergers.) This means that numerical solutions for waveforms obtained from alternative gravity theories would produce a large family of possible templates to compare with the data. When a full numerical solution of the MOG field equations is obtained, this one free parameter will be determined according to the strong gravitational field input of the merging black holes. Nonetheless, the fitting of the predicted MOG templates can produce different masses and spins for the merging black holes while still fitting the observational data. For example, MOG can predict lower component masses for the binary black hole mergers. This means that, using the LIGO data for the current detected events, and many more in the future, it is difficult to decide which gravity theory is correct. Potentially, astrophysical investigations of inspiraling compact objects such as black holes and neutron stars will help to distinguish general relativity from alternative gravity theories such as MOG.

OTHER WAYS OF DETECTING GRAVITATIONAL WAVES

LIGO is not the only experiment existing or planned to detect gravitational waves. There are three other proposed ways of detecting gravitational waves: the tracking of pulsars, laser interferometry in space, and detection of radiation from the early universe. We can observe radio, infrared, ultraviolet, and X-rays in addition to visible light. This fact allows us to make important new discoveries about the universe, including

gravitational waves, which we now know travel at the same speed as electromagnetic radiation.

Recall that the first indirect detection of gravitational waves came from
the timing of signals from pulsars orbiting one another. An astrophysicist, Donald Backer, at the University of California at Berkeley, proposed
in 1982 that one could use fast-spinning, millisecond pulsars to detect
gravitational waves by measuring small differences in the pulsar signal
arrival times at Earth. Gravitational waves from pulsars arriving at Earth
would push the planet a tiny bit toward some pulsars and away from
others. Backer envisioned a large pulsar array with Earth at the center of
the pulsar web. The more pulsars that could be detected, the bigger the
shift of Earth as gravitational waves passed through it. Backer's suggestion
would mean the gravitational wave amplitude or strength would be much
larger than those detected by LIGO. However, from the point of view of
everyday considerations, the strength of the gravitational wave is still very
weak—meaning, it would take years of pulsar measurements to detect a
gravitational wave. The measurement of pulsar timing signals is one of the
most accurate measurements in physics. They constitute what we call a *ce-
lestial clock*. But at the time Backer made his proposal, not enough pulsars
had been discovered to make this idea viable.

Later, in 2004, astrophysicist Richard Manchester of the Commonwealth
Scientific and Industrial Research Organization, began using the 64-meter-
diameter Parks radio telescope in Australia to find and track the timing
of pulsars in the southern hemisphere. At about this time, astrophysicist
Scott Ransom at the National Radio Astronomy Observatory and Andrea
Lommen at Franklin and Marshall University both located in Lancaster,
Pennsylvania, and Fredrick Jenet at the University of Texas at Brownsville
collaborated to form the North American Nanohertz Observatory for
Gravitational Waves (NANOGrav). With this project they launched a serious way to use the tracking of the pulsar signals to detect gravitational
waves. They use the 100-meter Green Bank telescope in West Virginia and
the 300-meter Arecibo telescope in Puerto Rico, which are the world's
two most powerful radio telescopes. These huge radio telescope dishes

can detect faint radio waves and capture pulsar signals with a timing as accurate as the best clocks on Earth. So far, not enough pulsars have been identified to claim that gravitational waves have been discovered by the pulsar array tracking system. Nonetheless, these astrophysicists hope that the Square Kilometre Array, a very large radio telescope array in Australia and South Africa, will have success in detecting gravitational waves when it begins operating.

At about the same time as the NANOGrav project and LIGO were launched, physicists proposed a LIGO-like experiment to be placed in space: LISA. The project was originally a joint NASA–European Space Agency (ESA) endeavor, which was supposed to have been launched by 2020. However, because of the high cost of the project, NASA pulled out in 2012. The ESA continued supporting the project, and now it is called eLISA, indicating it is strictly a European project. Because the interferometer system is to be in space, the interferometer arms can be built very long and, in fact, will reach 1 million kilometers! The eLISA project is planned to be operating in 2034.

In December 2016, ESA launched the LISA Pathfinder mission, with the intent of demonstrating that the project was technologically feasible. The Chinese plan to launch a satellite interferometry project, TianQin, in the future. As already described, the LISA project promises to detect very-low-micro-Hertz gravitational wave frequencies—that is, very long wavelengths—which would permit the detection of gravitational waves emitted by supermassive black holes orbiting one another. Recall that the frequencies detected by LIGO of gravitational waves from merging black hole binaries are in the range of 30 to 500 Hertz.

Yet another way of detecting gravitational waves is to observe the primordial gravitational waves emitted by the Big Bang about 14 billion years ago. On March 17, 2014, physicists at Harvard University and other institutions claimed to have detected primordial gravitational waves. They used the South Pole BICEP2 telescope to detect a particular signature in the polarization of the gravitational wave imprint on the polarization observations associated with the CMB. Recall that the claim was subsequently proved to be wrong, and the supposed detection was an effect of interstellar foreground dust in the Milky Way. New projects being considered to detect primordial gravitational waves will hopefully eventually overcome the problem of foreground dust. One of these projects is called the Cosmology Large Angular Scale Surveyor, or CLASS, being built in the Atacama Desert in northern Chile. Such projects will be able to measure many wave lengths, which can permit them to distinguish between background radiation emitted by dust and true gravitational radiation.

On October 3, 2017, Barry Barish, Kip Thorne, and Rainer Weiss (Figure 7.6) were awarded the Nobel Prize for their founding work leading to the discovery of gravitational waves. Unfortunately, one of the founders, Ronald Drever, passed away seven months before the announcement.

Figure 7.6. LIGO Nobel Prize winners (left to right): Rainer Weiss, Kip Thorne, and Barry Barish. Credit: LIGO

8

LIGO/Virgo Listens
to Neutron Stars

The detection of faraway merging black holes has been an unqualified success for Big Science. Yet all along, physicists at the LIGO collaboration have hoped to detect the merging of a black hole and a neutron star or binary neutron stars, for this would significantly increase our understanding of the physics of gravitational waves. Moreover, a neutron star–neutron star merger that was closer to Earth than the merging black holes would provide the opportunity for a simultaneous optical detection of the event. Such a simultaneous gravitational wave and optical source detection would herald a new and dramatic era in astronomy and cosmology.

FIRST BINARY NEUTRON STAR MERGER

On August 8, 2017, at 12:41 UTC (Coordinated Universal Time), the LIGO/Virgo observatories detected a strong burst of gravitational waves. Just 1.7 seconds later, the Fermi Gamma Ray Space Telescope detected a flash in the southern sky. What the astronomical observers had detected was a short gamma ray burst (GRB). During the subsequent hours, the observers realized they had seen the short GRB at the same time that LIGO was detecting the emission of gravitational waves from the collision of two neutron stars. This is a dramatic event, with far-reaching consequences for the future of astronomy. It heralded the beginning of so-called *multimessenger astronomy*. All important observations in astronomy up until this time were based on electromagnetic wave sources spanning the electromagnetic spectrum. Now a new cosmic messenger had arrived on the scene—namely, gravitational waves—which the astronomical community has now named a gravitational *siren*.

The source of this remarkable gravitational wave and electromagnetic wave emission was only 130 million light years away in an elliptical galaxy called NGC 4993, located in the constellation Hydra.

The discovery of this event was announced on October 16, 2017, in Washington, DC, at a press conference held jointly by the LIGO and Virgo projects, and organized by the National Science Foundation (NSF). This was the first time that LIGO was joined by the European project in making an announcement. Also about this time, 50 articles from the collaboration were announced investigating the astrophysical features of the event. Seventy telescopes in space and on the ground succeeded in collecting data over the entire electromagnetic spectrum during the weeks following the event. This dramatic event also had the far-reaching consequence of demonstrating that GRBs could be produced by colliding neutron stars.

THE GRB DETECTION

GRBs are very energetic explosions observed in distant galaxies. These explosions are the brightest electromagnetic sources of energy known to occur in the universe. The bursts can last from several milliseconds to hours. After the initial flash of short-wavelength gamma rays, a longer duration of afterglow is emitted as long wavelengths, such as X-ray, ultraviolet, optical, infrared, microwave, and radio waves.

There are two possible ways that GRBs are produced. One is the electromagnetic energy released during a supernova, when a massive star undergoing rapid rotation collapses to form a neutron star or a black hole. A second class of GRBs is thought to originate from the merger of binary neutron stars. The massive tidal forces experienced by the binary neutron stars can cause a precursor burst observed as a short GRB. This precursor burst is caused by the interaction between the neutron star cores and their crust resulting from the very strong tidal forces between the neutron stars seconds before their collision. All known GRBs originate in galaxies outside our Milky Way.

The combined event of gravitational wave and electromagnetic wave emission is called GW170817/GRB170817A. It is the first binary neutron star merger detection in the LIGO/Virgo collaboration.

WHAT HAPPENED WHEN?

The neutron star collision timeline can be described as follows: At around zero seconds, gravitational waves from two inspiraling neutron stars begin to appear in the LIGO and Virgo data, and within a fraction of a second, the two neutron stars merge (Figure 8.1). Two seconds later, the Fermi satellite detects a GRB optically. Then, about 14 seconds after that, the Fermi satellite sends out an automated message of detection. About 6 to 10 minutes later, the LIGO/Virgo observatories' software identifies the gravitational wave signal as originating at the source of the GRB. The astronomy community is notified about this gravitational wave detection about 40 minutes later.

In total, LIGO received about 100 seconds of signal from GW170817 in the data, accumulating a strong signal-to-noise ratio over this time.

Escaping neutrinos are expected to be detected from a neutron star merger, and the first neutrino detection comes in one hour later from the IceCube neutrino observatory at the South Pole. An accurate map of the source direction of the gravitational wave event is obtained five hours later from the LIGO/Virgo data. Eleven hours after time zero, the Swope Telescope in Chile reports a first optical detection and is able to pinpoint the host galaxy of the neutron star merger. As many as five other observatories independently obtain an optical image within an hour of the Swope Telescope. An ultraviolet emission from the source is detected 15 hours

Figure 8.1. Artist's depiction of two merging neutron stars, showing the gamma ray burst as narrow beams shooting out. Credit: National Science Foundation/LIGO/Sonoma State University/A. Simonnet

later by the Swift satellite. The ground-based Magellan Telescope, in Chile, measures the first optical spectrum of the neutron star merger event. As many as nine days later the Chandra space satellite detects X-rays, and 15 days later the Very Large Array observatory in New Mexico detects the emission of radio waves. Clearly, the merging of two neutron stars does not happen instantly, as it does in a black hole merger, when no optical measurements are involved.

It is interesting that the highest energy photons and neutrinos with relativistic energies were not detected as emissions from the source. Astronomers had expected that the collision would produce a very-high-energy natural particle accelerator. But no such high-energy particles were detected, and consequently, the accelerator hypothesis remains just that.

WHERE DO THE HEAVY ELEMENTS COME FROM?

The discovery of the neutron star merger has major implications for astrophysics. It offers the first evidence that heavy metals such as gold, platinum, and uranium were created by the merger explosion. Astrophysicists had thought that the heavy elements were mainly produced by supernovae explosions, but now what is called the *rapid neutron capture process*, or *r-process*, shows that merging neutron stars mainly create the heavy elements.

The r-process consists of neutron stars eating up free neutrons before they can undergo radioactive decay. These free neutrons are produced by the collision of the two neutron stars. In a matter of seconds, the neutron capture mechanism, or r-process, creates large amounts of elements heavier than iron. These heavier nuclei are eventually broken down by fission into alpha and beta particles, which produce a glowing thermal spectrum. (Alpha particles are helium nuclei, consisting of two protons and two neutrons, and beta particles are electrons.) The explosion resulting from the colliding neutron stars is called a *kilonova*, and it provides the best evidence, through the r-process, that nearly all of the universe's gold and platinum is produced by neutron star mergers. During the coming year, astrophysicists will be investigating to what extent other core collapse events, such as supernovae, play a role in the r-process.

One of the unexpected features of the kilonova event was that the intensity of the gamma rays emitted was orders of magnitude smaller than what was expected of a GRB at a distance of 130 million light years. Indeed, astronomers had to wait nine days to see X-rays being emitted, and the expected emission of radio waves was not detected before September 2, more than two weeks later. The suspected reason for this was that the gamma ray jet that shoots out from the merging neutron stars, which is perpendicular to the orbital plane of the progenitor neutron stars, was not aligned with Earth. The neutron star mergers emit gravitational waves in all directions, so that the gamma rays confined to a narrow jet should not be expected to be routinely aligned favorably enough to produce detectable GRBs. The LIGO/Virgo collaboration was incredibly lucky in that the first GRB event they observed from a binary neutron star merger was aligned favorably enough to produce a detectable GRB.

The tidal forces that disrupt the neutron stars leave their mark on the gravitational wave emissions. When the orbital frequency of the binaries reaches about 50 Hertz, the tidal forces diminish the energy of the final merger. These tidal forces and their effects on the neutron stars can tell us about the composition of the neutron stars, which consist of tightly packed nuclear matter. From a relativistic analysis of the tidal force strength, the LIGO/Virgo collaboration was able to determine the neutron star radii to be about 14 kilometers each. Once the neutron stars merged, the total mass of the resulting remnant was estimated to be 2.74 solar masses. This could be either a heavy neutron star or a very light black hole. Using general relativity, it has been estimated that a neutron star can be, at most, 3 solar masses before it collapses to a black hole. The most massive neutron star observed has a mass of 2 solar masses, so that if the merged neutron stars produce a heavy neutron star, this would be the heaviest neutron star ever observed. On the other hand, no one has observed a black hole with a mass as light as 2.74 solar masses. The issue of what the remnant of the neutron star merger is remains an open question, which may only be resolved when many more neutron star mergers are detected.

WHAT'S INSIDE A NEUTRON STAR?

The merging of two neutron stars and the emitted gravitational waves can tell us about what happens to the cores of neutron stars. Recall that the density of neutron stars is very high—about 100 trillion times greater than water. This enormous density breaks down the star's original atoms into neutrons, protons, electrons, and neutrinos. The protons and neutrons together form more neutrons and neutrinos. Moreover, because of the density of the neutron star core, the neutrons are broken down into their basic constituents, the up and down quarks that make up protons and neutrons.[1]

---

1. For a complete discussion of the quarks in the standard model of particle physics, see, for example, my book *Cracking the Particle Code of the Universe: The Hunt for the Higgs Boson* (2014).

The emerging "quark star" is not stable. However, the up and down quarks will also produce the heavier strange quark, because it takes just a little bit of energy to convert an up or down quark into a strange quark.

The process of producing strange quarks can go on indefinitely and can eventually turn the neutron star into a "strange star." The question then arises: how can we study the neutron star's or the strange star's interior? This is when the emission of gravitational waves produced by the merging neutron stars comes into play, because gravitational waves are emitted from the deep interior of the colliding stars.

So far, all of this is speculative. But, if a neutron star and a strange star collide, the LIGO/Virgo observatories could tell the difference between two ordinary neutron stars colliding. The reason is that a colliding system containing one strange star produces gravitational waves at a higher frequency than a collision of two neutron stars. Strange stars are believed to be smaller and denser than neutron stars, so when they spiral in to merge, the spiraling orbits shrink. This allows the compact objects to circle more rapidly, producing stronger and higher frequency gravitational wave emissions. The findings of research on the inner cores of neutron stars could settle whether strange stars exist, and whether they form the most stable matter in the universe. Indeed, like King Midas, this strange matter would turn all other kinds of matter it touches into strange matter.

MEASUREMENT OF THE HUBBLE CONSTANT

Another big new topic for research has emerged from the neutron star merger. By determining the distance from Earth to the merger, and combining that with the redshift of the host galaxy, it is possible to estimate the Hubble constant, which determines the rate of expansion of the universe through Hubble's Law, $v = H_0 D$, where v is the recessional velocity of any galaxy, H is Hubble's constant, and D is the distance to any galaxy. From the data analyzed, it was possible to obtain an initial estimate of H_0 to be between 62 kilometers per second per megaparsec (km/s/Mpc) and

82 km/s/Mpc. This is an important result because it allows astronomers to determine the Hubble constant independently of earlier methods used.

Currently, there are two conventional methods for determining the Hubble constant. One is a local measurement based on parallax measurements and using pulsating Cepheid variables[2] and measurements of supernova type 1 explosive events. These measurements determine the local value of H_0 to be about 73 km/s/Mpc. These measurements using the pulsating Cepheids and supernovae explosions refer to distances of about 30 megaparsecs. The analysis of the data has been performed by Adam Riess and his collaborators at Johns Hopkins University in Baltimore.[3]

Another measurement of the Hubble constant uses the CMB data for the early universe, which were obtained by the Planck satellite data collected in 2015. From an estimate of the cosmological distance to the CMB and the surface of last scattering in the very early universe, together with a measurement of the size of the matter fluctuations or ripples in the CMB, a value of H_0 is found of about 67 km/s/Mpc. The fact that the local measurement of about 73 km/s/Mpc is accurate to 2.3 percent, whereas the very distant CMB measurement is accurate to 1 percent, has stimulated controversy. Standard cosmology defines the Hubble constant to, indeed, be "constant," corresponding to a constant expansion rate of the universe. Adam Riess claims this tension will not go away, and it threatens the correctness of the standard model. Cosmologists hope that the LIGO/Virgo determination of the Hubble constant, based on new detections of neutron star mergers and GRBs, will produce an accurate enough value of H_0 to pin down the Hubble constant and potentially resolve the controversy.

2. Cepheid stars are regularly pulsating red giant stars that have been used as cosmic yardsticks since the early 20th century to tell the distance of the Cepheids from Earth.

3. A.G. Riess et al., "Large Magellanic Cloud Cepheid Standards Provide a 1% Foundation for the Determination of the Hubble Constant and Stronger Evidence for Physics beyond ΛCDM," *Astrophysical Journal*, **876**, 85 (2019).

HOW FAST DO GRAVITATIONAL WAVES MOVE?

A very important result from the analysis of the neutron star merger and almost simultaneous GRB is an extremely accurate determination of the speed of gravitational waves. The gravitational waves detected by LIGO/ Virgo arrived at the detectors 1.7 seconds earlier than the detection of the gamma rays. Either the gamma rays were emitted from the colliding neutron stars after the gravitational waves had left, the electromagnetic rays were delayed by interacting with intergalactic matter on the way, or a combination of both. This means that knowing the speed of electromagnetic waves, or photons—the difference between the gravitational wave speed divided by the electromagnetic wave speed minus one—is about 1 part in 10^{15}. From this we are assured that gravitational waves move at the speed of light, or electromagnetic waves, as they travel through the large cosmological distances of 130 million light years.

This result has significant ramifications for gravitational theory. A consequence of this result, of gravitational waves moving at the speed of light, is that it puts a severe restraint on modifications of Einstein's general relativity. Many proposed alternative gravitational models are falsified by this result. We pursue this important development in Chapter 9, on alternative theories of gravity.

9

Alternative Gravitational Theories

Physicists have come up with alternative theories of gravity ever since Einstein published his theory of general relativity, which was a modification of Newtonian gravity. Shortly after Einstein published his theory, Herman Weyl attempted to unify gravitation and electromagnetism in one geometric formulation.[1] Einstein himself began attempting to construct a unified theory of gravity and electromagnetism during the 1920s. These theories aimed to combine Maxwell's electromagnetic field equations with Einstein's gravitational field equations in a more general way than originally done by Einstein in his paper of 1916.

These attempts at unification of gravity and electromagnetism have not been considered successful, and with the discovery of new forces of interaction between particles, such as the strong force and weak force in the modern standard model of particle physics, it became clear that just attempting to unify gravity and electromagnetism could not lead to a complete unified theory. Maxwell's unification of electricity and magnetism in his field equations does represent a true unification of these two forces. But, it turned out that attempting to unify gravity and electromagnetism within one geometric scheme was not successful.

1. H. Weyl, "Gravitation und Elektrizität," *Sitzungsberichte der Preussischen Akad. D. Wissenschaften*, 465 (1918).

WHY SEEK ALTERNATIVE THEORIES OF GRAVITY?

There have been many proposed modifications of gravity over the years. Einstein's gravitational theory has a minimal mathematical content able to explain how spacetime is warped by matter–energy density. Any modification of this theory, which fits the observational data of the solar system extremely well, can result in mathematical inconsistencies, physical inconsistencies, and disagreement with observational data. To modify the accepted gravitational laws, one must have strong motivation provided by observations and data.

The standard model of cosmology, called the *Concordance Model* or the *Lambda Cold Dark Matter* (ΛCDM) *model*, describes cosmological data very well in terms of six parameters, of which the matter content is mainly "dark matter" and the cosmological constant. The standard model is based on general relativity with the addition of dark matter and dark energy to allow the theory to fit the data. However, because there is a lack of direct evidence for dark matter, in that there has been no successful experimental detection of dark matter particles, and there has not been any direct observation of the cause of the accelerating expansion of the universe, dark matter and dark energy remain controversial parts of the standard cosmological model.

Indeed, they can be considered "epicycles," which occur in the Ptolemaic model of astronomy invented by Greco-Roman astronomer Claudius Ptolemy during the second century CE. When the Ptolemaic model began to fail as a result of the astronomical observations of Danish astronomer Tycho Brahe and the seminal investigations of Johannes Kepler, it was necessary to invoke mending procedures such as epicycles and quants to the Ptolemaic model to make it fit the new data. The overwhelming dead weight of these additional hypotheses led to the revolution in astronomy that overturned the geocentric model of the universe. Is it possible that a resolution of the problems of dark matter and dark energy will lead to a new revolution in gravity, astronomy, and cosmology?

As far as the dark matter problem is concerned, a modification of Einstein's gravitational field equations requires that the alternative theory be consistent with certain observational data. Any alternative theory should predict the correct orbits of the planets, including the perihelion precession of Mercury, the bending of light by the Sun, and the time delay observations such as those obtained by the Cassini satellite probe. Moreover, it should agree with the observational confirmation of the equivalence principle, which states that bodies fall in a homogeneous gravitational field at the same rate independent of their composition. More to the point, the theory must explain the dynamics of galaxies—in particular, the rotation curves of stars orbiting inside them—without postulating dark matter to strengthen gravity. An alternative theory must also explain the stability of clusters of galaxies without the need for exotic dark matter. Moreover, it must accommodate the wealth of cosmological observational data.

The modifications of gravity proposed to explain dark energy and the accelerating expansion of the universe have to agree with the remarkably good fitting of the data based on the use of the cosmological constant in the standard cosmological model. These fittings with the data include an agreement with the calculations of the intensity of the fluctuations or the power spectra produced by the fluctuations observed in the CMB about 380,000 years after the Big Bang.

Large-scale surveys of galaxies—in particular, the statistical correlation of galaxies—lead to a determination of the "matter power spectrum," which is produced by the statistical counting of pairs of galaxies separated by any given angle on the sky. Counting pairs of galaxies leads to an estimate of the amount of putative dark matter that shows itself as dark matter "haloes" surrounding galaxies. The ΛCDM model with dark matter can fit the matter power spectrum data.

Can a modified gravity theory fit the matter power spectrum data without dark matter? Answering this question is tantamount to a critical test of the existence of dark matter in galaxies in the current universe. My modified gravity theory, MOG, can fit the currently available matter power spectrum data without dark matter. The crucial part of the test of whether MOG can fit the data is the appearance of predicted oscillations in the matter power spectrum. These predicted oscillations are produced by ordinary baryonic matter without dark matter. The baryonic matter interacts with photons before the surface of last scattering, when light is finally freed from its interactions with electrons and this pressure produces acoustical oscillations observed in the data.

As the number of pairs of galaxies in large-scale surveys increases, MOG predicts that the oscillations should also increase significantly. On the other hand, if dark matter exists in galaxies, then the dark matter will prevent the oscillations from growing. Ongoing large-scale surveys of galaxies such as the new Dark Energy Survey project, called *DES*, will increase the number of galaxies observed in the universe. Eventually, if dark matter exists in the current universe, and could be detected on Earth in laboratory experiments, the appearance of these oscillations will not become larger. The appearance or absence of these unit oscillations in the matter power spectrum data is a critical test of dark matter.

In addition to the motivation of wanting to unify all the forces of nature, physicists have desired to build alternative theories of gravity to accommodate cosmological principles such as Mach's principle, as well as new cosmological observations. Mach's principle states that the local inertial frame of reference and the force of inertia on a body are caused by the gravitational influence of all the distant matter in the universe. A gravity theory that includes this principle was proposed by Carl Brans and Robert Dicke in 1961, in which they attempted to interpret Einstein's gravity theory in terms of Ernst Mach's principle.[2] So far, the Brans-Dicke gravity theory, which is one of the simplest modifications of Einstein gravity, has neither been proved nor disproved experimentally. This theory has stimulated much theoretical research in gravitation theory.

In addition to explaining Mach's Principle, the Brans-Dicke gravity theory allows for a variation of Newton's gravitational constant with space and time. Paul Dirac was the first to propose that the gravitational

2. C. Brans and R.H. Dicke, "Mach's Principle and a Relativistic Theory of Gravitation," *Physical Review*, **124**, 925–935 (1961).

constant does vary on cosmological scales.[3] He called his idea the "large numbers hypothesis."

All of these demands on a modified gravity theory present a daunting task for any theorist who wishes to go down this path. There have been several modifications of gravity proposed that aim to obviate the need for exotic dark matter.

DARK MATTER

Although the standard model of cosmology, developed over several decades of research, is remarkably successful in describing the huge amount of observational data derived from telescopes such as the Hubble Space Telescope, dark matter and dark energy are still two serious fundamental issues that are not understood. Enormous effort has been put into understanding these two conundrums.

Swiss physicist Fritz Zwicky (Figure 9.1) started off the dark matter mystery when. in the early 1930s, he discovered that clusters of galaxies could not be stable without the gravitation from a large amount of dark matter in addition to the known ordinary matter formed from protons, neutrons, and electrons.[4]

There appeared to be not enough normal matter to create the gravitational attraction necessary to hold the galaxies together in clusters.

The need for dark matter in astronomy was strengthened during the late 1960s and the 1970s by astronomers Vera Rubin (Figure 9.2) and her collaborator Kent Ford, who discovered that the stars orbiting galaxies in circular orbits were moving at speeds significantly greater than could be explained by Newtonian and Einstein gravity.[5] These speeds are technically

3. P.A.M. Dirac, "The Cosmological Constants," *Nature*, **139**, 323 (1937).

4. F. Zwicky, "Die Rotvershiebung von Extragalaktischen Nebein," *Helvetica Physika Acta*, **6**, 110–127 (1933).

5. V. Rubin and W. Kent Ford Jr., "Rotation of the Andromeda Nebula from Spectroscopic Survey of Emission Regions," *Astrophysical Journal*, **159**, 379 (1970).

Figure 9.1. Swiss astronomer Fritz Zwicky. Credit: Wikimedia/Swiss Physical Society

Figure 9.2. American astronomer Vera Rubin. Credit: Carnegie Institution of Washington/AP

known as the *rotational velocity curves of stars*. To achieve agreement with the data, Rubin and Ford joined Zwicky in claiming that about 86 percent of the matter in galaxies must consist of dark matter. The problem is that this dark matter has stubbornly resisted being detected directly as dark matter particles in the current universe. Its existence is only inferred through its gravitational effects in galaxies and clusters of galaxies, and in cosmology.

We have learned that dark matter was postulated to exist to explain the strengthening of gravity in the anomalous dynamics of galaxies and clusters of galaxies. Some theorists believe the unsuccessful experimental efforts to detect dark matter particles directly (beyond dark matter's effect on gravity) is reason enough to search for alternative gravity theories.

There are several large, international experiments underway using liquid xenon to try to detect dark matter particles. The most popular candidate for the particles is the WIMP. This particle falls in the category of "cold dark matter" because the particle motion is nonrelativistic and thereby has little kinetic energy in the form of heat. Two recent experiments searching for WIMPs are Xenon1T, containing 1 ton of inert liquid xenon and located at Gran Sasso, Italy; and the PandaX-II, using a half ton of liquid xenon and located in China. The idea is to detect the recoil of dark matter particles colliding with xenon nuclei. The experiments are conducted deep underground to avoid contamination by cosmic rays emanating from deep space. They use xenon because it is a dense, inert liquid, which makes it more likely that incoming particles will hit the xenon nuclei. Because cosmic rays have been filtered out, dark matter particles would be the only incoming particles to hit the xenon nuclei. But so far, after decades of observations, no dark matter particles have been found.

In addition to underground experiments trying to detect dark matter particles, there are also satellite experiments that aim to detect the annihilation of dark matter WIMPs in our galaxy and beyond. These experiments use satellite telescopes that would be sensitive to large emissions of gamma rays and positron–electron pairs, which are postulated to be produced by the annihilation of the dark matter WIMPs. It is theorized that there should be large amounts of dark matter at the center of our galaxy. Indeed, a significant flux of gamma rays coming from the center of our galaxy has been detected by the Fermi laboratory space telescope.

But, new experiments strongly indicate that these gamma rays are emitted by old stars, such as millions of pulsars located in the bulge at the center of our galaxy. It is known that pulsars do produce significant amounts of gamma rays, and a million of these stars would produce a smooth spectrum of gamma rays, which could be misidentified as evidence for dark matter. In addition, the Fermi Gamma Ray Space Telescope has also detected a significant flux of electron–positron pairs, which again could be produced by the annihilation of dark matter WIMPs. However, as with the gamma ray emissions from the center of the galaxy, these electron–positron pairs are likely the result of known astrophysical mechanisms.

Another popular candidate for dark matter particles is the axion, which is an ultralight particle. The axion was proposed by Helen Quinn and Roberto Peccei during the late 1970s to explain an anomaly in quantum chromodynamics (QCD)—the theory of the strong interaction force binding quarks and gluons.[6] In technical terms, this anomaly is associated with charge-parity (CP) violation, which is a fundamental symmetry of particles in the standard model of particle physics. The axion is needed to explain why the theoretical CP violation in QCD does not agree with the experimental value, which is very small. A search for these ultralight axion particles has been ongoing for two or three decades with no success.

6. R.D. Peccei and H.R. Quinn, "CP Conservation in the Presence of Pseudoparticles," *Physical Review Letters*, **38**, 1440–1443 (1977).

DARK ENERGY

The idea of dark energy became a significant part of the standard cosmological model when astronomers Saul Perlmutter, Brian Schmidt, and Adam Riess published their data in 1998, claiming that observations of supernovae showed that their light was dimmer than it "should be," considering their distance from us as measured by telescopes.[7] The dimming of the supernova light suggested that the expansion of the universe was accelerating, rather than decelerating since the Big Bang, as was expected up to this time. This revolutionary discovery won the Nobel Prize for the three astronomers in 2011.

To explain the apparent acceleration of the expansion of the universe, it was necessary to come up with some kind of repulsive force that could counteract the attractive force of gravity in the universe. This hypothetical repulsive force was dubbed *dark energy*. This was an unfortunate choice of words because, according to Einstein's special relativity, energy is equivalent to mass when mass is multiplied by the square of the velocity of light. Yet the nature of dark matter and dark energy is different. Dark matter is believed to clump to form astrophysical objects, whereas dark energy seems to be smoothly distributed in space and, as described by the cosmological constant, has a uniform, constant density.

Theorists seized upon Einstein's cosmological constant as an explanation for dark energy. Recall, Einstein introduced the cosmological constant in his first paper on cosmology, published in 1917. At the time, the universe was pictured as being static. This was before Friedmann's discovery that Einstein's field equations predicted a dynamically expanding universe, and before Edwin Hubble discovered that galaxies were moving away from us and from each other as a result of the expansion of the universe. The effect of the cosmological constant was to produce a negative

---

7. A. Riess et al., "Observational Evidence from Supernovae for an Accelerating Universe and a Cosmological Constant," *Astronomical Journal*, **116**, 1009–1038 (1998).

pressure to counteract the expansion, and in effect to keep the universe static. This negative pressure came about through the suggestion by Zel'dovich during the 1960s that the cosmological constant was produced by vacuum energy with a negative pressure.

Thus, in 1998, Einstein's cosmological constant took center stage again, after having been dismissed by Einstein and others decades before. As often happens in theoretical physics, there was a fly in the ointment. A big one. When the energy associated with the vacuum mechanism producing the cosmological constant in quantum physics was calculated, it was fine-tuned to the ridiculous degree of one part in 10^{122}. The theoretical fine-tuning comes about because there is an upper bound on the vacuum energy in cosmology. Above this bound, the universe would collapse instantly. This result produced a deep suspicion about the whole suggestion that the acceleration of the universe was caused by Einstein's cosmological constant. This "mother of all fine-tuning problems" remains a serious issue in theoretical physics. This is part of what is called the *naturalness problem* in the standard model of particle physics and cosmology, and it remains unsolved. The naturalness means that in predictions of theoretical physics, there should not occur ridiculously large cancelations of numbers to arrive at the predicted value of a physical phenomenon.

DARK PROBLEMS

The invention of dark matter and dark energy to make the prevailing theories work has, as noted earlier, been pejoratively compared with the way ancient Greek philosophers added so-called epicycles to Ptolemy's widely accepted Earth-centered astronomy to fit the motions of the planets that had been observed. Recall that epicycles were simply invented to make the theory work, without any observational or theoretical underpinning. All of this raises the question: is Einstein and Newtonian gravity perhaps not correct? Could it be that there is no detectable dark matter in the current universe, but rather that the laws of gravity are not correct? Moreover, could it be that dark energy and the acceleration of the universe are an illusion?

The purported repulsive force responsible for the accelerated expansion of the universe—dark energy— represents another mystery, along with dark matter, that is needed to explain the cosmological data in the standard model. Dark matter is dark because we cannot directly measure its existence through electromagnetic observations. It is dark because we cannot directly detect it. The existence of both dark matter and dark energy is inferred only from gravitation using Einstein's general relativity. Despite this, there continue to be ongoing, large-scale experimental efforts devoted to detecting dark matter in underground experiments: at the LHC accelerator at CERN, Geneva, and using satellites.

The mystery of dark energy has produced a plethora of papers in journals attempting to explain its existence. These explanations take the form of modifications of Einstein's field equations. In general relativity, the cosmological constant is indeed a constant, whereas alternative gravity theories produce a time-dependent cosmological constant or a nonconstant source of repulsive force describing dark energy. If observations could show the cosmological constant is not constant, this would be a major boost for such alternative gravity theories.

Recently, the large-scale experimental project DES has produced observations that, with evermore precision, demonstrate the cosmological constant is a constant. An important signature of the veracity of the cosmological constant as a measure of the accelerated expansion of the universe is the equation of state describing the vacuum density of the universe. This equation states that the ratio of the vacuum density to the pressure of the universe is –1. The DES experiments to date, as well as the Planck satellite mission of 2015, have verified that the former ratio is –1 within a small margin of error. This indicates that the cosmological constant in the standard model can—at a phenomenological level—explain the accelerated expansion of the universe.

Despite this observational success, physicists are still not satisfied with the cosmological constant explanation for dark energy. On the other hand, they are not satisfied with modified gravity theories as an explanation of the fundamental mystery of dark energy either. There are, however, several well-known alternative gravity theories that can potentially explain the data without dark matter.

MODIFIED NEWTONIAN DYNAMICS (MOND)

In 1983, Israeli physicist Mordecai Milgrom published an article in which he modified the Newtonian acceleration law to fit the observational data for the speed of rotation of stars in galaxies.[8] We have seen that Einstein gravity can only explain these anomalous speeds by postulating dark matter. Milgrom's model is called MOND, which stands for modified Newtonian dynamics.

Milgrom's MOND model is a nonrelativistic modification of Newtonian dynamics—in particular, the gravitational acceleration law for particle motion. Technically speaking, it is a nonlinear modification of the Poisson equation describing Newtonian dynamics for weak gravitational fields. A critical acceleration parameter called a_0 is introduced in the MOND acceleration equation. By fitting the modified acceleration law to galaxy rotation curves, a reasonably good fit is obtained with the value a_0 equal to 1.2×10^{-10} meters per second squared. Above this critical acceleration, the normal Newtonian acceleration law is valid. Below the critical acceleration, Milgrom's modified acceleration takes over. It is then possible to fit the model to the observed rotational velocity curves of stars at the outer edges of galaxies. This model implies there is an acceleration threshold at which the behavior of gravity changes. The MOND critical acceleration a_0 plays a significant role in modeling galaxy dynamics.

Although Milgrom's model has been successful in fitting galaxy dynamics, it has not succeeded in modeling the observations of clusters of galaxies without having to add significant amounts of dark matter

8. M. Milgrom, "A Modification of the Newtonian Dynamics as a Possible Alternative to the Hidden Mass Hypothesis," *Astrophysical Journal*, **270**, 365–370 (1983).

to the model. Moreover, MOND is not successful in fitting the large amount of observational data describing the large-scale structure of the universe, such as the CMB. A significant criticism of MOND is that it has not been successfully shown that it can be compatible with Einstein gravity.

Although MOND was successful in describing galaxy dynamics, the theory was not fully compatible with Einstein's relativistic gravity. The MOND model is only valid for nonrelativistic motion of particles. That is, it is not a fully covariant theory of gravitation valid in every reference frame, as is the case with Einstein gravity.

Despite these criticisms and problems with MOND, as a result of its simplicity as a phenomenological model, compared to other possible modifications of gravity, it has been popular with astronomers. To circumvent the problems of fitting clusters of galaxies, various authors have proposed hybrid models using MOND to fit galaxies without dark matter, and simultaneously fitting cluster dynamics with dark matter. A criticism of that approach is that once you have been forced to include directly observable dark matter, both in cluster dynamics and cosmology, you have admitted that dark matter exists. Why, then, consider modifying Einstein's gravity theory?

ALTERNATIVE THEORIES CONSISTENT WITH EINSTEIN GRAVITY

Several theorists have attempted to develop alternative theories that are compatible with Einstein gravity. Again, although Maxwell's theory of electromagnetism was combined with Einstein's theory during the early development of general relativity, physicists did not succeed in forming a geometric unification of gravity and electromagnetism. As we recall, Hermann Weyl attempted to unify gravity with electromagnetism in 1918, but his efforts were not accepted by the physics community, and especially not by Einstein.

Jacob Bekenstein published a paper in 2004 promoting an alternative theory that was made to be consistent with Einstein's gravity theory.[9] The theory is called Tensor–Vector–Scalar Gravity (TeVeS). Beyond the normal single-metric tensor of Einstein gravity, TeVeS couples another metric tensor to matter. Thus, this is a bimetric theory. In addition, extra vector fields are included in the theory. One of these vector fields leads to the theory having a special frame of reference. In other words, the theory violates Einstein's special relativity. The choice of an arbitrary function and several parameters allow for TeVeS to reduce to the modified acceleration law of MOND for weak gravitational fields and slowly moving particles. This again introduces the same problem, that the theory cannot explain cluster dynamics and cosmology without dark matter. However, because of Bekenstein's earlier work on black hole thermodynamics in conjunction with Hawking radiation, physicists were predisposed to view his new theory positively. Many physicists have published papers on Bekenstein's TeVeS since 2004.

The experimental verification revealed by gravitational wave detection from the neutron star merger and its optical counterpart, that gravitational waves move at the speed of light, has falsified Bekenstein's TeVeS. TeVeS has two different metrics: one that describes the interaction of photons with matter, and another that describes the interaction of gravity with matter. The two metrics in the theory call into question that gravitational waves move at the speed of light, as they do in general relativity. Currently, there is no known satisfactory relativistic extension of MOND's nonrelativistic acceleration formula that avoids the falsification by the experimental evidence that gravitational waves move at the speed of light. Several other modifications of Einstein gravity have all met the same fate as TeVeS.

9. J.D. Bekenstein, "Relativistic Gravitation Theory for Modified Newtonian Dynamics Paradigm," *Physical Review*, **D70**, 083509 (2004).

Another attempt to discover an alternative gravity theory consistent with general relativity is my generalized gravitational theory called *Scalar–Tensor–Vector–Gravity* (STVG), or as it is popularly known, MOG ("modified gravity" or "Moffat gravity").[10] With MOG, I resolve the problem of dark matter by explaining the observational data for galaxy and cluster dynamics. The MOG theory explains without dark matter how gravity is strengthened at galactic size scales; it successfully fits the data in the absence of detectable dark matter. The basic reason for this agreement with data is that MOG strengthens the gravitational interaction with ordinary baryonic matter, without the need for dark matter. It does this by increasing the strength of the gravitational coupling constant. The theory can explain the merging of colliding clusters of galaxies, such as the so-called Bullet Cluster and the Train-Wreck Cluster, identified in the cluster catalogs as Abell 520, without the need for exotic dark matter. The predictions of the theory also agree with solar system observations.

In contrast to other generally covariant modifications of Einstein gravity, MOG agrees with the neutron star merger experimental result that gravitational waves move at the speed of light. The MOG theory has only one metric, as in the case of Einstein gravity, whereby gravitational waves and electromagnetic waves are predicted to move at the same speed. Another important feature of MOG gravity is that the equations of motion of massive particles or bodies lead to the particles falling at the same rate in freefall, independent of their material composition. This agrees with the equivalence principle. Therefore, the theory succeeds in retaining an important and well-verified prediction of Einstein gravity. However, the freely falling particles do not move along geodesics, as is the case in Einstein's gravity theory.

Let us consider in more detail the construction of STVG or MOG. The MOG modification of Einstein gravity consists of adding new gravitational degrees of freedom. In addition to the Einstein metric tensor,

10. J.W. Moffat, "Scalar-Tensor-Vector Gravity Theory," *Journal of Cosmology and Astroparticle Physics*, 0603:004 (2006).

the theory also contains a vector field that produces a repulsive gravitational force and two scalar fields. A principal scalar field allows the gravitational "constant" to vary in space and time in a manner similar to the tensor–scalar theory of Jordan-Brans-Dicke. In the language of particle physics, the metric tensor describes in quantum language a massless spin-2 particle that causes an attractive gravitational force, whereas the vector field describes a massive spin-1 particle (a spin-1 massive graviton), which causes a repulsive gravitational force, and a massless spin-0 scalar particle that causes an attractive gravitational force. All three of these particles are bosons. There is another scalar field describing the effective mass of the spin-1 particle. These additional degrees of freedom form a general gravitational theory in four-dimensional spacetime based on second-order partial differential equations.

In contrast, general relativity is based only on the massless spin-2 boson (graviton) described by the metric tensor in four-dimensional spacetime. If we take into account Einstein's cosmological constant, which also occurs in MOG, then there are two repulsive, antigravity mechanisms in MOG. The addition of an enhanced gravitational strength through the spin-0 scalar field, together with the spin-1 graviton particle describing repulsive gravity, allows MOG to fit a large amount of gravitational observational data. For example, this permits the fitting of galaxy and galaxy cluster dynamics without detectable dark matter, and also allows for the fitting of the significant amount of cosmological data, including the CMB data from the very early universe.

To distinguish MOG from general relativity—that is, to tell which theory is the better description of nature—we need critical observational data from strong gravitational fields. The recently detected gravitational waves will afford a valuable possibility of making such a distinction between the theories, through analyzing the currently available data and future gravitational wave detections.

The observations of the Event Horizon Telescope also provide a rich source of testing black hole physics in the presence of strong gravitational fields. Until now, all tests of gravitational theories, including Einstein's general relativity, have been for weak gravitational fields such as those in the solar system, in galaxies, and in clusters of galaxies. The gravitational fields in cosmology are also weak, except at or near the Big Bang.

THE MOG ALTERNATIVE TO BLACK HOLES

The MOG theory field equations lead to exact static, spherically symmetric solutions in the absence of matter. They also yield exact solutions for rotating black holes, which extend the Kerr spinning black hole solution of Einstein gravity. These solutions describe generalized Schwarzschild-MOG and Kerr-MOG black hole solutions. The spacetime metric for the rotating MOG black hole contains three parameters: the mass (M), the angular momentum (J), and the MOG parameter alpha (α). These parameters have to be measured by observing black holes. When α is equal to zero, the metric solution reduces to the one obtained from Einstein gravity. These solutions of the MOG field equations open up the possibility for MOG to explain the nature of compact astrophysical objects such as neutron stars and black holes.

Recall the no-hair theorem of Wheeler, which states that the only "hair" a black hole can have is its mass, spin, and electric charge. The electric charge of astrophysical bodies is negligible because these bodies instantly neutralize themselves. This will also be true for black holes. A negligible electrical charge would be neutralized in a black hole because the positive and negative electric charges would annihilate one another. In addition, recall that because the Coulomb electrical force is 10^{40} times bigger than the gravitational force, any significant electrical charge on a black hole would make the black hole blow up. As is the case in Einstein gravity, the MOG black holes are characterized by only the mass and spin of the black hole, not the electric charge. However, the predicted consequences of the MOG black holes can be significantly different from the predictions of general relativity black holes.

One important consequence of the MOG equations for dark compact objects is the prediction for gravitational waves produced by coalescing binary black holes or neutron stars. In MOG, as in general relativity, the gravitational waves move at the speed of light. In MOG, while still agreeing with the LIGO data of calculated gravitational waveforms during the inspiraling phase, the masses of the binary black holes and the final mass of the quiescent black hole resulting from the merging of the two initial black holes can be different from the masses predicted by Einstein's gravity theory. This is because the strength of the Newtonian gravitational coupling constant is enhanced for MOG black holes.

Indeed, in the formulas describing the coalescence of the binary black holes, mass is always multiplied by the stronger gravitational constant. A bigger gravitational constant requires a smaller mass to fit the data. This prediction can have important consequences for the evolutionary explanation of binary black holes forming through stellar collapse and supernovae explosions. Recall that from the observations of X-ray binaries, consisting of progenitor stars and black holes orbiting each other, the estimated masses of the black holes are about 10 solar masses or less, significantly lighter than the more than 30 solar masses obtained from the LIGO data with a gravitational constant equal to the Newtonian gravitational constant. It is possible that MOG can lower the component masses of the binary black holes to be more in agreement with the 10-solar mass black holes in X-ray binaries.

The LIGO measurements of the effective spin parameters of the merging black holes—which are, on average, zero or negative for the detected events—can be accommodated by MOG, which retains the alignment of the spins expected in some binary black hole evolutionary models. Recall that the spins of the black holes in X-ray binaries in our galaxy prefer to be aligned with the orbital angular momentum of the system, in contrast to the spins of the black holes in other galaxies more than a billion light years away from Earth. As we have already asked: why should the black holes in distant galaxies act differently from those in our galaxy? This disturbing question has not yet been answered.

For the MOG solution of the field equations without matter, the mass enters as an integration constant, as in the case of the static Schwarzschild solution in general relativity. The solution depends on one free parameter, called *alpha*. A Kerr-MOG solution for rotating black holes including spin is also obtained from the field equations. These two solutions represent the end state of the gravitational collapse of a massive progenitor star to an ultracompact object with a mass bigger than a neutron star.

My basic MOG paper also contains another class of exact solutions of the MOG field equations. For these solutions, if the parameter alpha (α) is bigger than a critical value, α_{crit}, equal to 0.673, then the final state of the collapsed object has neither event horizons nor a singularity at the center. It is a MOG "dark star." If, on the other hand, α is less than α_{crit}, then the object is a black hole with two concentric horizons, but no singularity at the center. If, in nature, there is a physical mechanism that demands that α is greater than α_{crit}, then MOG would be predicting only the existence of a regular dark star solution without a central singularity or an event horizon.

For either the MOG dark star or the MOG black hole, astronauts falling into these objects will not be spaghetticized by *infinite* tidal forces because there is no internal singularity. However, there are still strong tidal forces present. A 3-solar mass black hole will produce a tidal acceleration amounting to nearly a billion g's—a huge acceleration for an average-size human. Moreover, in contrast to the black hole Schwarzschild solution of Einstein's gravity theory, the redshift of light at the surface of the MOG dark star is large and finite. In Einstein gravity, on the other hand, the redshift is infinite at the event horizon.

As described, a gravastar is a solution of Einstein's field equations, bringing in quantum mechanics to remove the event horizon of a black hole general relativity solution. The MOG dark star is a classical solution of the MOG field equations, not incorporating quantum mechanics. It does not have event horizons or a singularity at its center. The redshift of a photon falling through the MOG dark star is finite but sufficiently large that it looks like a black hole from the outside. Near the center of the MOG dark star, the matter can be interpreted as being a Bose-Einstein condensate, as in the case of the gravastar. Alternatively, it can be interpreted as an effective vacuum state. The rest of the matter in the MOG dark star resides between the singularity-free center and the horizonless surface. The object is stable and the core of the dark star, consisting of the Bose-Einstein condensate fluid, makes it different from a stable neutron star. In a MOG neutron star, the equation of state is based on standard strong interactions and the nuclear physics of neutrons, with degenerate neutron gas that provides the pressure that balances attractive gravity and makes it a stable object.

With increased and more accurate LIGO data, it will be possible to calculate the frequencies of "quasi-normal modes" from the vibrational ringing of the final remnant MOG dark star, which was created by the merging of two horizonless MOG dark stars. One can then use these calculations to distinguish between Einstein gravity black holes and MOG dark stars. The LIGO data are tested against a quarter of a million templates of numerical solutions of Einstein's field equations for the merging of black holes or neutron stars. The large number of potential numerical relativity calculations solving MOG's field equations will be different from the general relativity ones. The masses and spins of the merging Einstein gravity black holes and MOG dark stars will be different. These differences will require a reinvestigation of the new numerical relativity templates obtained from solving MOG's field equations.

A fruitful way to test the difference between general relativity and MOG black holes with horizons from the horizonless MOG dark star is to measure the properties of light waves passing through these three objects. The light waves can propagate through the singularity-free MOG dark star and be partially reflected by the ring of light barrier outside the Schwarzschild radius of the dark star, and this partially reflected light will then go through the dark star and again be partially reflected by the light ring barrier. This bouncing back and forth of reflected light is not possible in an Einstein gravity black hole. The partial reflection will create a ping when it hits the ring of light barrier, producing an echo that can be observed in the ring-down phase frequency data of the LIGO gravitational wave event detections. The black holes with event horizons will not produce these echoes, but will just show a fading away of the vibrational ringing of the final black hole. It may take many gravitational wave detections before enough data are available to resolve this test of whether black holes with horizons exist.

As with a gravastar, the MOG dark star could potentially resolve paradoxes such as the information loss paradox, and the other peculiar paradoxes of black holes. If the matter inside the horizonless dark star is sufficiently viscous, then the waves passing through the dark star will be damped, and no echoes will be seen. Moreover, because the dark star is horizonless, it is not expected that it will evaporate through Hawking radiation. Although significant quantum physics is needed in creating a gravastar, this is not the case in the creation of a classical MOG dark star. It remains to be seen whether this is a virtue in the future of theoretical physics and experimental investigations distinguishing gravastars, MOG dark stars, and black holes. Clearly, at this time, we do not have any convincing observational evidence that can distinguish among these possibilities.

THE STRING ALTERNATIVES

String theory has produced many alternative theoretical ideas in particle physics and cosmology since the early 1970s. The theory is based on the hypothesis of higher dimensions, with the strings described not as zero-dimensional points in spacetime, but as one-dimensional strings that vibrate like the string on a violin. But so far, no experimental evidence has come forth verifying the existence of more than three spatial dimensions and one time dimension. Nonetheless, physicists have been actively pursuing the idea that particles are one-dimensional vibrating strings in 10 or 11 dimensions.

Unfortunately, as of now, there has been no definite experimental test of string theory. Experiments at the LHC at CERN have not confirmed the existence of any extra dimensions above the three spatial dimensions we live in. Many string theorists have developed ideas about the effects string theory might have for black holes, ranging from solutions for black holes that do not have central singularities and so-called "fuzzy" black hole event horizons, which are put forth to solve Hawking's information loss problem.

The LHC has also failed to detect the existence of supersymmetric particles. The postulated supersymmetry of spacetime plays an integral role in the formulation of superstring theory.[11] Moreover, research during the past decade has revealed that superstring theories have an enormous number of vacuum states—in fact, about 10^{500}. These are potential vacuum solutions of the string field equations, which are called the *landscape* of string theory. The existence of this landscape makes it very difficult for the theory to make any definite predictions that can be compared to experiments. It has led to the idea of a "multiverse," a large or infinite number of universes of which ours is putatively one. By its nature, this multiverse theory cannot be tested experimentally, because we are unable to observe any universe outside of our own. However, speculative ideas have been put forward by several eminent physicists, claiming it is possible to detect the signatures of these other universes. It has even been suggested that a signature of another universe can be detected in the CMB data. So far, this proposal has not been validated.

HIGHER SPATIAL DIMENSIONS CONSTRAINTS ON MODIFIED GRAVITY

Another constraint on modified gravity can be obtained from the gravitational wave and optical multimessenger event of the merging neutron stars. Some alternative gravity theories predict a difference in speed between the propagation of electromagnetic waves and gravitational waves. The multimessenger neutron star data have ruled out such theories.

---

11. The term *superstring* implies that string theory incorporates supersymmetry of particles. In fact, string theory has to include supersymmetry to be consistent with the fermion particles of the standard model.

The extradimensional theories of gravity predict there is a leakage of gravitational waves into the dimensions higher than the four dimensions of spacetime of general relativity. In general relativity, the amplitude, or "strain," of gravitational waves diminishes inversely proportionate to the luminosity distance to the source of gravitational waves. In higher dimensional gravity theories, the expected leakage of gravity would further weaken the gravitational wave amplitude. For black holes with masses 10 to 100 times the mass of the Sun, gravitational wavelengths are about a hundred to a thousand kilometers, which are in the range that LIGO is most sensitive to. These gravitational wave measurements can constrain higher dimensional models.

In an article by Kris Pardo, Maya Fishbach, Daniel Holz, and David Spergel, the authors discover, by analyzing the combined gravitational wave and optical data of the neutron star merger, that the spatial dimensions are constrained to be three, which is consistent with the 3 + 1 spacetime dimensions of general relativity.[12] In higher dimensional theories, physicists have invoked screening mechanisms to suppress the amount of leakage into higher dimensions. Analysis of the data also significantly constrains the screening mechanism, and the conclusion is that the best fit to the data confirms the four-dimensional spacetime of general relativity. This analysis of the gravitational wave data does not constrain string theory in any way, because the 10-dimensional string theory "curls up" the extra spatial dimensions into tiny objects about the size of the Planck length—10^{-33} centimeters—which is well beyond the capability of gravitational wave measurements. Published papers have claimed to rule out the Dvali-Gabadadze-Porrati model using the gravitational wave data. However, the Randall-Sundrum model is probably not ruled out because the gravitational waves were not able to penetrate the fifth dimension sufficiently in this model.

12. K. Pardo et al., "Limits on the Number of Spacetime Dimensions from GW170817," *Journal of Cosmology and Astroparticle Physics* (2018).

THE HOLOGRAM ALTERNATIVE

The idea of a hologram for black holes and cosmology was first proposed by Gerard 'tHooft.[13] Leonard Susskind provided a string theory interpretation.[14] It has been proposed that string theory allows a lower dimensional space description in which gravity emerges as a hologram. For the case of a black hole, all the three-dimensional information contained in the volume of the black hole is projected onto the two-dimensional surface of the event horizon, as a hologram. String theorists such as Argentinian physicist Juan Maldacena and other authors claim that the holographic principle resolves the black hole information loss paradox because all the information within the black hole is preserved on its event horizon, and allows a loss of information.[15] The information is not trapped within the black hole, but can escape from the event horizon surface. However, this solution to the information loss problem is controversial and not fully understood within a complete quantum gravity formulation.

The idea of the holographic principle has also been applied to cosmology, where all the information inside the universe is projected onto the cosmological horizon. This idea has not been fully established as being valid because the cosmological horizon is time dependent, because of the expansion of the universe.

The hologram principle applied to black holes leads to a thermodynamic interpretation of black hole entropy, as formulated by Bekenstein and Hawking. Here, the black hole entropy is determined by the surface area of the event horizon and not by the volume of the black hole, as would be the case in a normal macroscopic body.

13. G. 't Hooft, "Dimensional Reduction in Quantum Gravity," *arXiv gr-qc*/9310026 (1993).

14. L. Susskind, "The World as a Hologram," *Journal of Mathematical Physics*, **36**, 6377–6396 (1995).

15. J.M. Maldacena, "Eternal Black Holes in AdS," *Journal of High Energy Physics*, **2003**, 021 (2003).

The hologram principle idea has stimulated many publications. But, as with the multiverse idea, it can only be considered speculative at best. As with many modern physics proposals, testing the hologram idea is beyond the reach of current physical experiment.

QUANTUM GRAVITY THEORIES

As with the electromagnetic field, and with matter, many physicists believe the gravitational field should also be treated as a quantum mechanical phenomenon. In the Einstein field equations, the left-hand side of the equations is interpreted as being geometric, whereas the right-hand side describes the components of matter. The right-hand side must be "quantized" to conform with the modern quantum mechanical description of matter. The argument then goes that the geometric spacetime description of the left-hand side of Einstein's field equations should also be treated quantum mechanically. Therefore, spacetime should be quantized as bits of quantum energy. The detection of gravitational waves by the LIGO observatory also reinforces the idea that gravity should be quantized. This is because all matter waves are quantized, like electromagnetic waves. Why should gravitational waves be treated any differently than the other waves of nature?

A successful quantum gravity theory might have significant repercussions for black holes. It may provide deeper explanations for the quantum physics active at the black hole event horizon, if indeed such an event horizon exists, and it might remove the troublesome singularity lurking at the center of the black hole. A quantum gravity theory may not mathematically allow for singular, dense points in spacetime, either at the cosmological Big Bang or at the center of black holes.

During the 1960s, Roger Penrose published a paper in which he proved that, given certain positive conditions on energy, if you contracted a classical black hole down to a point, then the point must inevitably become a singularity in space.[16] Subsequently, Hawking, and

---

16. R. Penrose, "Gravitational Collapse and Space-time Singularities," *Physical Review Letters*, **14**, 57 (1965).

later Hawking in collaboration with Penrose, proved that in a classical treatment of cosmology, when the universe contracted to a point at the Big Bang, this would produce a singularity in time.[17] The proof demanded positive energy conditions (i.e., the density of matter is always positive). I stress here that these Hawking-Penrose "singularity theorems" are based on purely classical physics. It seems unreasonable that if you contract a black hole to a point with an infinite density of matter, that quantum physics would not play a role in the subsequent fate of the black hole. In particular, quantum gravity may play a role in erasing the singularities.

The problem with quantum gravity, in its earlier iterations based on general relativity, is that its calculations led to unphysical infinities. The great success of the standard model of particle physics is that the quantum field theory on which the model is based leads to a renormalizable quantum field theory. When a quantum field theory is renormalizable, the infinities associated with the mass and charge of particles can be canceled out, leading to finite physical results such as the cross-section for the scattering of two particles. In other words, infinities that arise in calculations of particle interactions can be renormalized away, leading to finite results that can be compared to experimental data. Applications of quantum field theory to a quantum gravity theory lead again to infinities, but because of the nature of the quantum gravity field equations, these infinities cannot be renormalized away as in the standard model of particle physics. However, the goal of quantum gravity is to circumvent these infinities in the calculations by either improving the way Einstein's field equations are quantized or by generalizing Einstein's equations in such a way that all calculations lead to finite results.

Many attempts have been made over decades to resolve the problem of infinities in quantum gravity. One possible route was string theory. String theorists have claimed that the theory includes gravity. The string vibrations are identified with particles, and one such vibration is the graviton, which is the quantum equivalent of the photon in quantized electrodynamics.

---

17. S. Hawking and R. Penrose, *The Nature of Space and Time*, Princeton University Press, 1996.

Because of the one-dimensional nature of the string, quantum gravity calculations will produce finite results, which were not possible to obtain in the case of the zero-dimensional point graviton. Despite this promising finite result in string quantum gravity, there are no observational data for quantum gravity that can verify string quantum gravity theory.

Other possible quantum gravity theories have been proposed, such as nonlocal quantum gravity, in which the interactions between gravitons and matter particles are extended in the quantum field theory calculations in such a way that all the quantum gravity calculations are finite. I promoted this type of nonlocal quantum gravity theory during the late 1980s and early 1990s. I also extended this theory to the standard model of particle physics, which led again to a quantum field theory that was finite and physically self-consistent. However, in the case of the nonlocal quantum gravity theory, it was not possible to verify any finite calculations with experimental data. Frustratingly for the theorist, there are no quantum gravity data.

Why has it been impossible to verify quantum gravity? When the gravitational field is quantized, it predicts the existence of a graviton, which is the carrier of the gravitational force, the quantum package of energy equivalent to the photon energy carrier in quantized electromagnetic field theory. However, because of the extreme weakness of the gravitational force, it would take a high-energy accelerator the size of the galaxy to detect a graviton. Alternatively, a detector in a gravity experiment would have to be the size of the planet Jupiter to detect the existence of a single graviton. Physicists such as Freeman Dyson have claimed that it is impossible ever to detect gravitons or verify their existence.

Although we are faced with the lack of experimental detection of the graviton, we still must resolve the theoretical conundrum of Einstein's field equations—namely, that the geometry of spacetime on the left-hand side of the equation should logically match the quantum description of matter on the right-hand side of the equation. However, certain physicists argue that a semiclassical description of quantum gravity—in which the right-hand side of Einstein's equation is described by quantum matter, and the left-hand side can be described by classical geometry—provides an adequate description of quantum gravity.

As has been demonstrated through the history of physics, a physical theory can only be considered successful if it describes experimental data without being falsified—that is, if the results of calculations in the theory are not negated by experimental data. This is the catch-22 for quantum gravity—that we demand from theoretical physics that a quantum gravity theory exists, but so far we have not been able to justify the existence of the theory experimentally or falsify its predictions. It may be that we never will be able to obtain experimental quantum gravity data.

Then again, there is the even more radical view that the geometry of spacetime should not be quantized. Indeed, there is so far no evidence that spacetime should be quantized. Einstein was opposed to the idea, and he maintained this position until the end of his life. The future of quantum gravity will influence how we attempt to understand black holes and what goes on inside a black hole. Apart from whether we need quantum gravity, all matter has to be quantized, with the consequence that whatever the material content of a black hole, it will have to be treated quantum mechanically.

THE FIREWALL PROPOSAL

One recent speculative proposal by theorists Ahmed Almheiri, Donald Marolf, Joseph Polchinski, and James Sully brings in quantum physics in an attempt to resolve Hawking's information loss paradox.[18] They speculated that when an observer approaches the black hole event horizon, the particle physics active near the event horizon produces an extreme heating mechanism called a *firewall* that would incinerate a spacecraft and its astronauts. If this theoretical speculation were true, then human

18. A. Almheiri et al., "Black Holes: Complementarity or Firewalls?" *Journal of High Energy Physics*, **2013**, 62 (2013).

observers could never actually approach an event horizon, as seen by distant static observers, because they would be reduced to ashes. In effect, the incinerating of observers amounts to the destruction of any information ever going through the event horizon. Therefore, this resolves the information loss paradox by preventing information from going through the event horizon in the first place!

However, we have to revert to the relativistic time dependence of observers in general relativity. Observers falling freely through an event horizon do not detect the existence of the horizon and therefore are not incinerated by this quantum physics firewall phenomenon. Let us suppose that astronauts in a spaceship attempt to get close to a black hole horizon. To hover just near the horizon, the spacecraft must accelerate to counteract the immense gravity of the black hole. This would require an extremely large acceleration. We note that the surface gravity of a 10-solar mass black hole is estimated to be about a billion kilometers per second squared!

According to a proposal by Canadian physicist William Unruh, such an acceleration would produce a heat bath of radiation. For the extreme acceleration needed to keep the spaceship hovering near the horizon, the temperature of the radiation would be extremely high, which would—in the end—incinerate the spacecraft and its passengers. In addition, the acceleration needed to keep the spaceship hovering near the horizon would be so large that it would flatten the spaceship.

The extreme heat produces a kind of firewall near the horizon, without the direct involvement of quantum physics and particle physics. However, if the spacecraft is not subject to extreme acceleration to keep it in position near the event horizon, then it just freely falls through the horizon without any extreme temperature of radiation. In other words, it is possible to keep the classical general relativity explanation and the equivalence principle near the event horizon without the firewall. Recall that the equivalence principle states that all bodies fall in a gravitational field at an equal rate. The firewall proposal is speculative and controversial, and can only be considered as one of the many proposed resolutions of the information loss paradox.

FACTS AND FANTASIES

There are many possible, and published, resolutions of the Hawking information loss paradox. None of these has any experimental confirmation—a
fact related, of course, to the lack of any experimental confirmation of
the existence of Hawking radiation in the first place. Indeed, there are
physicists who propose there is no information loss paradox. The existence of so many speculative ideas about black holes only emphasizes that
modern physics is awash with such theories, and we have entered an epoch
of speculation akin to what existed in pre-Galileo and pre-Newtonian
times, or even akin to Aristotle's philosophical approach to science.

We hear a lot about "postfact" and "posttruth" in the politics of our
times. Much of the speculation engulfing modern fundamental theoretical physics can be described as postfactual. Two or more quite different
"truths" can exist in physics and be believed by their advocates, although
neither is empirically falsifiable. During the Middle Ages, astronomers
and mathematicians produced horoscopes based on the motions of the
planets and their positions in the zodiac. All the everyday decisions of
humans relied on these horoscope predictions to determine activities
in farming, wars, family life, and religion. Sometimes, the horoscope
predictions of astrology were correct; often, they weren't. A similar situation exists in modern physics, as a result of the often lack of firm experimental data that can separate false theories from correct ones. Today
we use economic theory to decide much of the business world's activities,
which can also be correct sometimes, and also often false. So from this
point of view, human theories that predict the future can be just as unre-
liable as astrology. Johannes Kepler—who discovered the three basic laws
of planetary motion, and that the planets move along elliptical orbits, and
that the heliocentric model of Copernicus was the correct astronomical
theory of the solar system—made a living producing horoscopes.

The history of science has shown us that for a physics paradigm to last,
and to meet the test of time, it must be supported by continuing experimental evidence. During the past five decades, there has been much speculation about the nature of black holes. It is necessary for observational

astronomy to weed out much of this theoretical speculation. Up until now, this has not been possible for the obvious reason that we cannot directly observe black holes as isolated entities. This has now changed with the observation of gravitational waves by the LIGO/Virgo collaboration. However, the LIGO/Virgo detection of gravitational waves from coalescing binary black holes is still an indirect observation of black holes. This is because these binary black holes are isolated and without any physical "clothing"—, they are "bare" or "naked" black holes without the typical accretion disks that occur, for example, in the stellar-mass X-ray binary black holes, and can be directly observed. The "observation" of the black holes only occurs through linking the observed waveforms detected by the LIGO experiments to theoretical calculations based on the assumption that general relativity is the correct theory for strong gravitational fields.

The new EHT experiments will further the possibility of direct observations of black holes by observing the shadow, or silhouette, of the black hole against the bright ring of photons, called the *photon sphere*, surrounding the black hole and attached to the outer accretion disk of gas. In this new era of observational astronomy of black holes, general relativity and all the theoretical alternatives are being put to the test.

10

The Biggest Eyes in the Sky: The EHT

Imagine if you could train your telescope on the huge black hole at the center of our Milky Way galaxy. Given that black holes are, by definition, invisible, what could you see? The astonishing aim of the Event Horizon Telescope (or EHT), also referred to as the Very Long Baseline Interferometry (or VLBI) project, is to observe the *shadow* of the supermassive black hole, Sagittarius A-star (or Sgr A*), at the center of our galaxy, almost 26,000 light years away, as it obscures the black hole's accretion disk. The project will also investigate the shadow of the supermassive black hole at the center of galaxy M87, which is 53.5 million light years away.

This shadow, or silhouette, is expected to show up in the data as a dark area where the black hole eclipses the bright emissions coming from the accretion disk and from distant light sources behind the black hole. This radiation has been trapped by the black hole's strong gravitational field. The invisible black hole will block the light in the form of photons coming from distant sources behind it, and will also block radiation coming from the part of the accretion disk that is behind the black hole, so that the invisible black hole will be seen as a silhouette on a background of light. Although we refer to light or photons coming from the accretion disk and the background of the black hole, the EHT project—because it is composed of an array of radio telescopes—only "sees" the radio wave part of the light spectrum. This is as close to observing a black hole directly as our technology allows today.

Observations of the silhouette or shadow will provide much more information about supermassive black holes than the deductions that have so far been made from the elliptical orbits of nearby stars, such as the S0-1 and S0-2 stars orbiting the Sgr A* black hole that have been observed. Their highly eccentric orbits have been accurately measured by telescopes making observations in the infrared part of the light spectrum. These astronomical observations have determined that the gravitating object Sgr A* is indeed supermassive, with a mass of about 4 million Suns, and exerts a powerful gravitational pull on nearby stars.

As well as studying the black hole at the center of our galaxy, the EHT will also be observing a supermassive black hole in the center of galaxy M87 (Messier87). This galaxy was one of a number of "nebulae" (later called *galaxies*) named after French astronomer Charles Messier, who catalogued them in 1751. M87, located in the constellation Virgo, is one of the most massive and nearest galaxies to the Milky Way.

Important information about the supermassive black hole at the center of our galaxy will also come from observations of radiation emitted by its accretion disk. Hopefully, this will tell us whether the event horizon exists and, if it does, its size and shape. Preliminary observations by the EHT and simulations of models of the accretion disk have shown that strong magnetic fields are present there. These magnetic fields are included in models to explain the intermittent flares of radiation observed at the accretion disk and large jets of matter that shoot out from the supermassive black hole in the galaxy M87.

THE SUPERMASSIVE BLACK HOLES AND ORBITING STARS

Two rival teams of astronomers have been aiming their telescopes at a star orbiting Sgr A* to confirm the existence of that monstrous black hole at the center of our galaxy. The star, called S0-2, is a young, blue star that follows an elongated orbit and streaks past the supermassive black hole every 16 years. Its orbit places it 18 billion kilometers away from the black hole at its closest approach. During these times, the black hole's strong gravity slows the vibration of light waves, stretching them into the red end of the electromagnetic spectrum. This gravitational redshift was one of the first predictions of Einstein's gravity theory, that a massive body shifts the vibrations of light waves to the red end of the spectrum. The observations of S0-2 provide a possibility of observing this redshift as a result of the strong gravitational field of the supermassive black hole.

The two collaborations that have been pursuing this project are using the Very Large Telescope Interferometer, an array of four giant telescopes in Chile. One is the international collaboration led by Reinhard Genzel at the Max Planck Institute for Extraterrestrial Physics. Genzel uses the array of telescopes, with infrared detectors, to observe hundreds of stars orbiting the blurry core of the center of the galaxy that is about one tenth of a light year across. The group has observed knots of gas orbiting the galactic center. The gas clouds circle the center of the galaxy about every 45 minutes, completing a circuit of 240 million kilometers at about 30 percent of the speed of light. The astrophysicists claim that only a supermassive black hole could fit within such a tiny orbit.

The other collaboration in the project is the European Southern Observatory, a multinational consortium with headquarters in Munich, led by Andrea Ghez, a professor at the University of California at Los Angeles. Dr. Ghez is looking for extreme astrophysics at the center of the galaxy. The two rival teams have been able to measure the gravitational redshift of star S0-2, and claim the redshift data agree with general relativity. This information provides a more accurate measure of the total mass of the monstrous black hole called Sagittarius A*, confirming that it is a supermassive black hole. In addition to this important observation, the teams have also detected evidence of hot spots, or flares, creating a tiny blur of heat at what must be the location of the supermassive black hole. When S0-2 comes close to the supermassive black hole again in the coming years, it is hoped that we will learn about other important physical properties of the black hole.

HOW THE EHT WORKS

The EHT project requires an extraordinary orchestration of submillimeter-wavelength radio telescopes on Earth. So far, about 60 institutions around the world, including universities, research institutes, and government agencies are collaborating on the project. The EHT ties together radio telescopes on mountaintops in places such as Arizona, Hawaii, Mexico, Chile, Spain, as well as Antarctica, to create, in effect, a planet-size radio telescope with a baseline that covers the whole western hemisphere of Earth (Figure 10.1).

To capture images of the black hole with sufficiently accurate resolution requires connecting all of these telescopes so they effectively operate as one large telescope. This is a huge technological undertaking. Picking an observation time is, in itself, difficult. In fact, the only time that the array of telescopes can be "turned on" to observe the giant black hole is

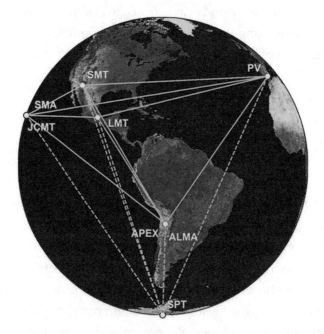

Figure 10.1. Planet-size telescope. The eight radio telescope stations over six geographic locations, which the Event Horizon Telescope project used in April 2017 to gather data on supermassive black holes. Credit: EHT Collaboration

for a couple of weeks in the spring of every year, when the weather allows observations from all the sites and when Earth is oriented correctly to probe the center of our galaxy. Earth must also have the right orientation to observe the elliptical galaxy M87, which is about 55 million light years away. Galaxy M87 has a mass of 10^{12} solar masses, and a supermassive black hole in its center that is 3 to 7 billion times the mass of our Sun. (The uncertainty in the determination of the mass is discussed later.)

In April 2017, eight telescopes in the array, in six locations, were linked together and turned on at the same time. Initial data were collected, but the team had to wait until December 2017 to receive data from the South Pole because the winter weather prevented any cargo, including the EHT computer disks, from leaving Antarctica. The data then went to supercomputers at the Massachusetts Institute of Technology (MIT) and at the Max Planck Institute in Bonn for analysis.

The EHT project's planet-size virtual telescope is the largest ever created. Because of its enormous size, the sharp resolution of the interferometry observations is about 1000 times better than the Hubble Space Telescope. Sheperd Doeleman, the head of the EHT collaboration, has said that the precision of the measurements of the black hole shadow is equivalent to someone holding up a quarter in Los Angeles and someone in New York being able to read the print on the coin.

The accretion disks surrounding the two target supermassive black holes are emitting X-rays, gamma rays, and radio waves. The data for the radio waves arrive as a classical wave front interference pattern, or "fringes," at each of the radio telescopes. Supercomputer algorithms are then used to put together the patterns from all the linked telescopes to construct a picture of the shadow of the Sagittarius A* black hole, which is 26,000 light years away from Earth. Although the supermassive black hole in M87 is 55 million light years away, its central black hole is about a thousand times more massive than Sgr A*. Even though it is about 1000 times farther away, its image will be only a little smaller when viewed from Earth than Sgr A*. An important goal for the EHT data analysis, as with the gravitational wave analyses by the LIGO/Virgo collaboration, is to determine whether Einstein's general relativity is indeed correct for strong gravitational fields or whether we need to modify the theory by using an alternative gravity theory. Of course, we also hope to learn a great deal more about black holes than we have learned in the past.

WHAT WILL THE BIGGEST EYES IN THE SKY SEE?

The accretion disk, formed over billions of years as the black hole sucks in dust, gas, and stellar material, surrounds the black hole. One of the challenges in obtaining an image of the Sgr A* black hole is for the EHT to penetrate the near part of the accretion disk cloud between the black hole and Earth to be able to observe the silhouette of the black hole inside the accretion disk.

As supermassive as the black hole is at the center of our Milky Way, its event horizon is expected to be only 24 million kilometers across, which is only about 17 times bigger than the Sun. The mass is 4 million times the mass of the Sun, and the density inside the black hole is about 138 times the density of water. So, Sgr A* is supermassive rather than superdense. Its accretion disk, consisting of turbulent clouds of gas and dust, emits huge amounts of energy in the form of radiation, and its matter is voraciously eaten by the black hole.

To solve the problem of "seeing" through the accretion disk cloud to the black hole's shadow, astronomers use wavelengths of light in the radio wave band. Compared to visible light, radio waves are not scattered significantly by the accretion disk matter. To choose a particular wavelength for the observations, thousands of computer simulations were performed, modeling the accretion disk gas. The outcome of these simulations was that the optimal wavelength should be 1.33 mm. This wavelength allows astronomers to see nearly to the black hole event horizon. Another positive outcome of this choice of wavelength is that Earth's atmosphere does not prevent the light in this part of the spectrum—that is, radio waves—from coming from the black hole to the telescope dishes on Earth. The image the astronomers finally see will look like a crescent. It has a crescent shape because the accretion disk gas is rotating around the black hole, so that the Doppler effect of gas moving toward us on Earth appears significantly brighter than the gas moving away from us, creating an image of a bright crescent attached to a dark disk.

WHAT WILL THE EHT DATA MEAN?

During the two years of anticipating the EHT data analysis and results, physicists and astronomers discussed the important experimental outcomes of the observations. The results will put Einstein's theory of general relativity on the line. The shadow image will be of a purely geometric origin, determined by the amount of curvature of spacetime created by the mass of the black hole. The size and shape of the shadow are precisely predicted by Einstein's theory of gravitation. After many computer simulations of the shadow, astrophysicists claimed that its size can be determined within a few microarcseconds.

If the measurements do not agree with the predictions of Einstein's gravitational theory, then this would be a dramatic and unexpected result. It would mean that Einstein's gravitational theory fails for strong gravitational fields. It would also mean that his theory has to be generalized. That is, an alternative gravity theory would need to take its place, after more than 100 years.

Although most physicists working on the EHT project expect Einstein's theory to be upheld by EHT measurements, some hope for a significant deviation from Einstein's gravitational predictions because this would be a dramatic discovery for the EHT collaboration, and would provide fodder for research projects for decades.

THE EHT ANNOUNCEMENT OF RESULTS

On April 10, 2019, the U.S. National Science Foundation (NSF) organized a press conference at the National Press Club in Washington, DC. The NSF was a major funder of the EHT project, along with the European

Space Agency (ESA). The Washington, DC, press conference was one of six simultaneous press conferences, the others being in locations where the radio telescopes were situated.

The excitement had been building up to the 9:00 a.m. EDT (Eastern Daylight Time) press conference, ever since the conference had been announced 10 days before. We had been waiting for the EHT results for two years, since April 2017, when eight of the radio telescopes had succeeded in obtaining data for the two supermassive black holes, Sgr A* and the one in the elliptical galaxy M87. It had taken this long for the data to be analyzed, and for conclusions to be reached. The excitement was palpable once the press conference got underway.

Shep Doeleman, the head of the EHT project, declared that the results from the galaxy M87 were the strongest evidence of the existence of black holes to date, and also the strongest evidence backing up Einstein's gravity theory. "We've seen what we thought was unseeable!" declared Doeleman.

My colleague from the Perimeter Institute, Avery Broderick, who had flown to Washington to be part of the NSF panel, said, "General relativity has passed another crucial test."

When asked by a reporter how the scientists felt about seeing the image of the M87 black hole (Figure 10.2) for the first time, Doeleman described the "relief and surprise" and the "astonishment and wonder" at seeing the "extraordinary image."

"It brought tears to my eyes!" added France Córdova, the head of the NSF, who is also an astrophysicist.

For an hour, members of the panel described how the EHT worked, described the image in more detail, and discussed the future plans of the project. It had been a challenge just preparing the individual telescopes to work together. Technicians had to install specialized hardware in the participating telescopes, such as atomic clocks and data-gathering equipment. That was particularly difficult in the harsh environment of Antarctica. The team had been extremely fortunate in April 2017, when the weather cooperated, and all eight telescopes were able to focus on their two targets.

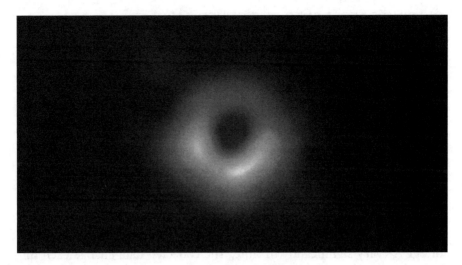

Figure 10.2. First picture of a black hole. The image of the M87* supermassive black hole, configured by computer algorithms, using the data collected by the EHT project in April 2017. Credit: EHT Collaboration

As for the data collected, the analysis will keep scientists busy for a long time. Dan Marrone, an experimental astrophysicist at the University of Arizona, explained that the data gathered in those few days in spring 2017 represent 5 petabytes, or a million billion bytes, which can be visualized as "a half ton of hard drives" or "the entire selfie collection over a lifetime for 40,000 people!" As for future plans for the project, more radio telescopes will be added, which will sharpen the resolution of the resulting images, and the team hopes to have at least one additional telescope in space.

When the panel talks finished, the floor was opened up to reporters. The first question, from an Associated Press reporter, voiced what had been on many of our minds while listening to the panel discussion.

"Have you captured images of Sagittarius A* [at the center of our Milky Way galaxy] yet?" Indeed, in the years leading up to this announcement, the EHT collaboration had emphasized Sgr A* as the centerpiece of the project, with M87 a secondary target. We had expected to hear a lot about Sagittarius A* at the press conference. Had something gone wrong?

DIGGING DEEPER INTO THE SUPERMASSIVE
BLACK HOLES

After the press conference, the *Astrophysical Journal* released six articles by the EHT collaboration explaining many more details about the project and its results. I was very interested in the lack of results for Sgr A*, and sought explanations in the articles.

We know that the mass of Sgr A* is about 1000 times smaller than the mass of M87*.[1] This means that the size of Sgr A* is also about 1000 times smaller. This has the consequence that the accretion disk rotates around Sgr A* much faster than the one surrounding M87*. So the images taken of Sgr A* vary significantly in time. The collected data change each minute, which makes it difficult to combine data from different times to construct a reliable image. From M87*, multiple data records can be stacked to improve the signal-to-noise ratio and allow construction of an image. However, it is difficult to construct an image of Sgr A* from the data using computer algorithms. To construct images from the Sgr A* data, one would have to film the object over a period of four nights, whereas for M87*, snapshots could be taken. For the computer algorithms to create images from the data, it is necessary for several images of the object to be essentially the same.

In addition to this problem, because Sgr A* is located at the center of the Milky Way Galaxy, foreground dust and gas obscure the black hole like a fog. This foreground scatters the radio waves, just like light, blurring the image. On the other hand, the radio waves coming from M87* do not appear to have to contend with this problem.

Although they are enthusiastic about the M87* (Powehi) results, the EHT collaboration will still be attempting to construct images of Sagittarius A*.

---

1. It has now become standard in the literature to refer the black hole in the M87 galaxy as *M87**—M87 star. It has also been called *Powehi*, after an 18th-century Hawaiian chant describing the creation of the Hawaiian universe. The word "po" in this chant means "embellished dark source of unending creation." The name honors the two radio telescopes in Hawaii that were part of the VLBI network.

SEEING THE UNSEEABLE

Most people who saw and read the news soon after the EHT announcement thought that the image of M87* was an actual photograph of the black hole. This is, of course, false. The image was constructed through an extensive process of analyzing the data by using computer algorithms.

The huge amount of data obtained by the virtual planet-size radio telescope came in as coordinated, time-synchronized radio wave observations made by eight telescopes at six different locations in the array: Arizona, Hawaii, Mexico, Chile, Spain, and Antarctica. In early April 2017, multiple three- to seven-minute scans were achieved. In addition to the observations of M87*, other objects were observed for the purposes of calibration. All the data were compiled on hard drives, with the observation times stamped on the drives from very accurate atomic clocks installed at each telescope station. The enormous amount of data was then combined, checked, calibrated, and then recalibrated many times. At this stage, the researchers had to turn the data into a static image. The collaboration had to decide how to proceed with this.

The project leaders divided those who were working on the imaging process into four groups, each of which would process the M87* data in different ways, creating their own images from computational algorithms. Two groups used a nearly half-century-old method, or algorithm, called *CLEAN*, whereas the other two groups used computer code employing a technique called *regularized maximum likelihood* (RML). Published papers developing the RML technique can be found as far back as 2007. It was widely used for constructing emission tomography images. It is based on a technique called *maximal likelihood estimate* (MLE).

Given a statistical model based on observations, the method estimates the parameters of the model by finding the parameter values that maximize the likelihood function. As an example, consider the heights of all adult females in France, even though we are unable to measure the height of every female in the French population. The model is based on the assumption that the heights of the females are a normal distribution with some unknown mean and variants. The mean and variants can be estimated with MLE, knowing the heights of some sample of the overall population. Taking the mean and variants as parameters, the goal is to find particular parametric values that make the observed results the most probable given the normal model.

In the case of the radio wave EHT data, the aggregated data is in the form of pixels, and MLE is applied to the pixels as parameters of the model. RML refers to a smoothing of the data when configuring the image. It is important to emphasize that the configuration of the data is claimed not to be dependent on a theory model such as general relativity. The important result is that the construction of the image displays a dark depression surrounded by light with a bright crescent on the bottom left of the image.

RML was developed further by Katie Bouman at MIT, Andrew Chael at Harvard, and other colleagues, for the purpose of analyzing the M87* black hole observations. They had to cancel out the effects of the atmosphere and piece together observational patches from each radio telescope's data (Figure 10.3).

The work of each group was kept secret. After several weeks of analysis, the four groups submitted their final results on a web portal. In June 2018, during a week-long meeting at Harvard, the image processing results were analyzed. The four groups first met separately, and then they began to share details. The final results were displayed as four images. Strikingly, the images all shared similar features, including a photon or light ring approximately 42 microarcseconds in diameter, surrounding a dark circular central region. The dark area was the shadow of the black hole's event horizon. Kazunori Akiyama, the team's coleader, told them they had made a monumental discovery. What is also exciting is that the image looked much like the simulated images obtained by Avery Broderick's group at the Perimeter Institute and the University of Waterloo. Inside the dark shadow of the already iconic image lurks the event horizon of the monstrous black hole.

A massive library of theoretical simulations of black holes had been developed by astrophysicists based on the assumption that general relativity was correct. To develop this library of simulated images, it was necessary to use a model called *general relativity magneto-hydrodynamic* (GRMHD)

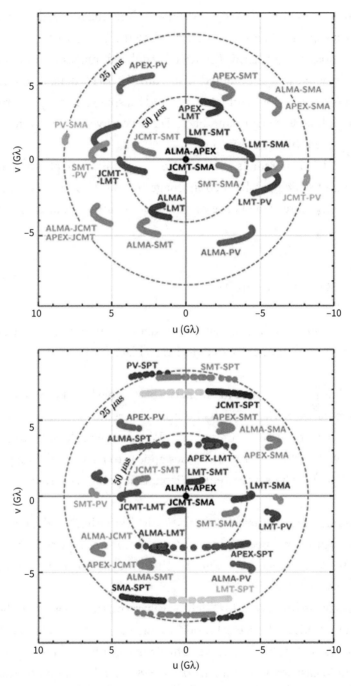

Figure 10.3. Aggregated data for M87*. Data collected by each Event Horizon Telescope (top panel) and, for calibration purposes, data from quasar 3C 279 (bottom panel), for the April 2017 observations. Credit: EHT Collaboration

simulations, which included a turbulent, hot, magnetized disk of gas orbiting a spinning black hole described by the Kerr spacetime metric. It was now possible to use this library of simulated images to compare with the constructed image obtained just from the EHT data. The GRMHD model produces a powerful jet of gas and plasma streaking out from the M87 black hole both front and back. For several years, optical telescopes have observed this enormous jet.

The model also predicts a shadow and an asymmetric photon emission ring, both of which can be seen in images of M87*.[2] The crescent seen in the EHT image represents the bright photon ring surrounding the dark shadow.

EVENT HORIZON OR NOT?

During the press conference, Avery Broderick maintained that the object at the heart of galaxy M87 *was* definitely a black hole.

> The object that powers M87's jets is a black hole like those described by general relativity . . . . If the object were not a hole, the image would have been very different. . . . We can now rule out a dim but otherwise visible surface. This does appear to have the defining feature of a black hole, the event horizon, that point of no return.

At the same time, Broderick expressed his desire that the data would reveal something unexpected. In answer to a reporter's question at the end of the press conference, he declared, "The most exciting thing we could do would be to supplant Einstein!"

Whether the results of M87* definitively prove there is an event horizon can be questioned. There exist models, called *exotic compact* or *dark compact objects* that do not have event horizons and have a kind of surface. They would cast shadows similar to the M87* shadow found by the EHT.

---

2. The photon emission ring does not necessarily coincide with the "innermost stable circular orbit" of photons, but is instead related to the gravitational lens photon ring.

It is possible to obtain from a modified gravity theory a classical solution that has no event horizon, has a surface, and does not have a singularity at its center, but appears to be very dark because of a high redshift at the surface, but not an infinite redshift, as is the case for an event horizon surrounding a black hole.

Other exotic compact objects are the gravastars and strange stars discussed earlier, boson stars,[3] and rapidly rotating compact objects that have a naked singularity at the center—that is, they do not have an event horizon. A naked singularity solution is a dark compact object without an event horizon, so that the singularity at the center is "visible." Recall, of course, that a singularity is an infinite density of matter at a point in space.

One of the observations obtained for M87*, and also similarly for Sgr A*, is that the flux of radiation decreases rapidly as one approaches the putative event horizon, to the extent that only about 10 percent of the expected radiation is emitted near what would be the event horizon. It has been argued by Broderick and Narayan that this is the signature of an event horizon, because the event horizon is sucking in up to 90 percent of the radiation. For a dark compact object, we would expect the surface to be "hard," like a star or a neutron star and, with a sufficiently high density of matter at the surface, it would emit a significant amount of radiation. This may be the case for a gravastar, but not for a boson star.

The problem with arriving at a definitive conclusion about this issue is that the arguments are model dependent, involving assumptions that currently cannot be proved or disproved. Models exist for which the surface of a compact object absorbs more radiation than it emits, and for M87* being a dark compact object without an event horizon. These models disagree with the conclusions of Broderick and Narayan. Ultimately, the radio

---

3. Stars and black holes are normally pictured as made of fermions, such as protons and electrons; a boson star is made of bosons such as the speculative ultralight axions. Strictly speaking, there is no observational evidence for their existence.

wave observations cannot penetrate sufficiently close to the putative event horizon to make a definitive conclusion.

Indeed, there is not unanimity within the EHT collaboration and their published articles about exactly what they have seen, exactly what is creating the shadow. An article published in *Astrophysical Journal Letters* by the EHT collaboration shortly after the press conference stated:

> [I]t is more difficult to rule out alternatives to black holes in [general relativity], because a shadow can be produced by any compact object with a spacetime characterized by unstable circular photon orbits. Indeed, while the Kerr metric remains a solution in some alternative theories of gravity, non-Kerr black hole solutions do exist in a variety of such modified theories.[4]

However, some exotic alternatives to black holes, such as naked singularities, boson stars, and gravastars, are admissible solutions within general relativity, and do provide possible models even though they may be considered contrived. The results from the M87* observations can already rule out the naked singularities solutions because the shadows they would produce are significantly smaller and very asymmetric compared to those of a Kerr black hole. In contrast, the M87* results for the shape of the shadow show that the shadow is circular, with a deviation from circularity of not more than about 10 percent. Also, certain exotic models based on wormholes predict much smaller shadows than have been measured. Boson stars, as models of compact objects with circular photon orbits, do not have a surface or an event horizon. However, the flow of the accretion disks for boson stars behaves differently from the Kerr black hole.

To sum up, the constraints obtained from the M87* observations on deviations from the Kerr geometry rely only on the validity of the

---

4. The Event Horizon Telescope Collaboration, "First M87 Event Horizon Telescope results: I. The shadow of the supermassive black hole," *The Astrophysical Journal Letters*, **875**, L1 (2019).

equivalence principle and are agnostic about which theory of gravity is ultimately being revealed by nature. However, with continued measurements with ever-improved precision, the parameters of the background space-time metric can be probed and potentially rule out alternative theories of gravity. In contrast, current gravitational wave observations of black hole mergers probe the dynamics of the underlying gravity theory but are not able to produce multiple and repeated measurements of the same dark compact source.

HOW MASSIVE IS M87*?

As I have discussed, one of the goals of the EHT collaboration is to determine whether Einstein's general relativity is correct for the strong gravitational field of black holes. In particular, does it predict the right size of the black hole shadow? The predicted angular diameter sizes of the light emission region and the shadow are proportional to the mass of M87*. Consequently, it is crucial that an accurate determination of the mass for M87* be obtained, so that it can be used to compare the size of the shadow and emission region with predictions of the assumed gravitational theory. The size of the shadow is measured as the "angular" size, which is the diameter of the shadow of the black hole divided by the distance to observers on Earth.

A number of estimates of the mass of M87* have been made, beginning in 1978, by Wallace Sargent and collaborators. To date, a large number of M87* black hole astrophysical estimates have been made, but there remain open questions. There are two common ways of using astrophysics to determine the mass of M87*. One depends on the stellar dynamics of the M87 galaxy and the other is based on models of the dynamics of the gas surrounding M87*. The stellar-dynamical method involves a complex model, called the *Schwarzschild model*, which determines the gravitational potential including the central black hole, a model of the bulges and disks of the galaxy, the dark matter halo, and one or many stellar components. The M87 galaxy is a giant elliptical galaxy and does not have a disk. The model can be biased based on a number of systematic effects, such as the mass-to-light ratio in the vicinity of the black hole and farther out in the galaxy, dark matter in the galaxy, and the intrinsic shape and orientation effects of the galaxy and the orbiting stars within the galaxy.

The other, gas-dynamical model is conceptually simple, but whether the gas motion is circular remains to be verified. The gas is assumed to be controlled by a Newtonian gravitational potential, as it is sufficiently far away from the event horizon of the black hole to ignore general relativistic effects.

There have been tests to judge the consistency between the stellar- and gas-dynamical methods of measuring the supermassive black hole masses. From the beginning, the fact that the M87 galaxy is one of the most luminous nearby galaxies and hosts one of the most massive black holes known has made it an important target for astrophysicists to study. So far, there have been estimates of black hole masses in other galaxies in which it was found that the stellar- and gas-dynamical mass measurements were consistent with one another. Other black hole mass measurements have been made that showed a discrepancy between the two models, including the supermassive black hole in M87.

In a 1997 paper by Fernando Macchetto and colleagues, the method of determining the black hole mass of M87* from the gas-dynamical model was described in detail.[5] An important source of data includes the atomic spectral lines of the gas obtained using the Hubble Space Telescope. This spectral line data and the velocity of the gas were fitted by a rotating thin-disk model. In other words, the model pictures the gas as an accretion disk spinning around the black hole, and estimates the mass of the black hole by observing how fast the disk is moving.

Another paper on the mass of M87* was published in 2013 by Jonelle Walsh et al., in which they created an improved map of the kinematic

---

5. F. Macchetto, A. Marconi, and D. J. Axon, "The Supermassive Black Hole of M87 and the Kinematics of Its Associated Gaseous Disk," *The Astrophysical Journal*, **489**, 579–600 (1997).

structure of the gas disk, using new, more accurate Hubble Space Telescope measurements of the spectral lines of the gas.[6] They were also able to determine a value for the inclination angle of the disk projected onto the plane of the sky. In addition, they improved the understanding of the internal dynamics of the gas disk.

The resulting mass of M87* determined by the gas-dynamic model of Walsh et al. is $M = (3.5^{+0.9}_{-0.7}) \times 10^9$ solar masses. This is similar to the earlier mass determined by Macchetto et al. also using the gas-dynamical model: $M = (3.2 \pm 0.9) \times 10^9$ solar masses. An earlier determination of the black hole mass by Richard Harms and collaborators in 1994 had resulted in a mass consistent with both the Macchetto et al. and Walsh results, but with a larger error.

Meanwhile, several authors since the 1970s have used stellar-dynamical models to determine the black hole mass in M87. In a paper by Karl Gebhardt and Jens Thomas in 2009, a determination of the M87* mass was obtained using the stellar-dynamical model.[7] The authors used high spatial resolution observations of stellar light in the central region of the M87 galaxy and globular cluster velocities inside the galaxy to determine the black hole's mass. They assumed there was a dark matter halo around M87 and assumed a value for the stellar mass-to-light ratio of the galaxy. The dark matter halo has to be considered, because those who believe in dark matter claim that all galaxies have dark matter halos. Gebhardt and Thomas argued that a dark matter halo was needed to measure the black hole mass reliably. The mass of the black hole is, in the end, determined by the velocities of the orbiting stars in the neighborhood of the black hole. A problem is that the M87 galaxy is about 54 million light years away, and telescopes cannot resolve individual stars at that distance. However, they

6. J. Walsh, A. Barth, L.C. Ho, and M. Sarzi, "The M87 Black Hole Mass from Gas-Dynamical Models of Space Telescope Imaging Spectrograph Observations," *The Astrophysical Journal*, **770**, 86–97 (2013).

7. K. Gebhardt and J. Thomas, "The Black Hole Mass, Stellar Mass-to-Light Ratio, and Dark Halo in M87," *The Astrophysical Journal*, **700**, 1690–1701 (2009).

can measure averages of orbits of stars from spectroscopic measurements of the stars.

In 2011, Gebhardt et al. published another paper that included more accurate spectrographic measurements obtained from the Gemini telescope.[8] This project consists of twin optical/infrared telescopes located in Chile and Hawaii that, together, can image the whole of the northern and southern skies. The data were combined with extensive stellar orbiting data out to large radii of the galaxy to derive a black hole mass equal to $M = (6.6 \pm 0.4) \times 10^9$ solar masses. The paper claimed that this mass determination is insensitive to the inclusion of a dark matter halo, and also to the mass-to-light ratio of the galaxy. This removed the model dependence on the dark matter halo and the mass-to-light ratio argued by Gebhardt and Thomas in 2009 to be a necessary part of the stellar-dynamical model.

However, there are still model assumptions that can significantly affect the measurement of the black hole mass. These include the assumption that we are observing the elliptical galaxy M87 edge-on, and that the galaxy is an axisymmetric and oblate spheroid. Elliptical galaxies have been observed that are triaxial in shape.[9] It is important to question the axisymmetric solution because a more general triaxial shape of M87 can introduce systematic errors leading to a significant difference in the mass determination. The estimate of the M87* mass from the stellar dynamical models can depend on assumptions about the M87 dark matter galaxy halo and mass-to-light ratios. The stellar- and gas-dynamical methods of measuring supermassive black hole masses can differ in their results by a factor of 2 or even 3, which is a huge discrepancy. For M87*, the discrepancy is approximately a factor of 2. Perhaps there are systematic errors that occur in both models that are much larger than the small errors claimed in the model calculations. Over the decades, the two ways of measuring

8. K. Gebhardt et al., "The Black Hole Mass in M87 from Gemini/NIFS Adaptive Optics Observations," *The Astrophysical Journal*, **729**, 119–132 (2011).

9. The equation of a triaxial ellipsoid centered at the origin with semiaxes a, b, and c aligned along the coordinate axes is $x^2/a^2 + y^2/b^2 + z^2/c^2 = 1$. The equation of a spheroid with z as the symmetry axis is $(x^2 + y^2)/a^2 + z^2/c^2 = 1$.

the supermassive black hole mass have resulted in the stellar-dynamical models being consistently higher for the mass than the gas-dynamical models.

WAS EINSTEIN RIGHT FOR STRONG GRAVITATIONAL FIELDS?

General relativity yields a prediction for the size of the shadow and the enclosing photon sphere or ring. The first published prediction that the black hole will cast a shadow against its photon sphere was made by John Lightman Synge in 1966.[10] Using the Schwarzschild solution describing a static, nonrotating black hole, he calculated the angular sizes of the shadow and the light emission region based on the trapping of photons by the strong gravitational field of the black hole. James Bardeen in 1973 extended the calculation of the angular sizes of the shadow and emission region to a rotating black hole using the Kerr solution to general relativity.[11] Astrophysicist Jean-Pierre Luminet published the first simulated image of a shadow cast by a black hole in 1979.[12] The dark depression in the image constructed by the EHT collaboration is identified with the shadow cast by the supermassive black hole M87*. The gravitational lensing caused by the bending of light by the strong gravitational field of the black hole, and the associated shadow, confirm that Einstein's gravitational theory correctly predicts these basic features of black holes. Certain alternative gravitational theories that have solutions of field equations that deviate from the Schwarzschild and Kerr solutions also predict gravitational lensing and a shadow, although they predict a different angular size of the shadow.

10. J.L. Synge, "The Escape of Photons from Gravitationally Intense Stars," *Monthly Notices of the Royal Astronomical Society*, **131**, 463 (1966).

11. J. Bardeen, "Timelike and Null Geodesics in the Kerr Metric," in *Black Holes [Les Astres Occlus]*, eds. C. Dewitt and B.S. Dewitt, New York: Gordon and Breach, 215–239 (1973).

12. J.P. Luminet, "Image of a Spherical Black Hole with Thin Accretion Disk," *Astronomy and Astrophysics*, **75**, 228 (1979).

The formulas derived by Synge, Bardeen, and Luminet for the angular size of the black hole showed that the size was proportional to the mass. This will be the case for the black hole M87*. The measured size of the shadow and the photon ring is 42 ± 3 microarcseconds. This value is determined by the analysis of the EHT data and the simulations of the shadow and photon ring based on assuming that the spacetime near and at the black hole is the Kerr metric solution of general relativity. The experimental resolution for the VLBI is resolved well enough to allow for the estimate of the angular diameter of the shadow and its photon ring. The prediction by general relativity of the shadow and photon ring angular diameter favors the higher mass of approximately $M = 6.6 \times 10^9$ solar masses, which is consistent with the black hole mass found from the stellar-dynamical model. The lower mass $M = 3.5 \times 10^9$ solar masses measured by the gas-dynamical model would predict an angular diameter of 22 microarcseconds, which is approximately half the value predicted by general relativity, and would not agree with the EHT-measured value. Some would conclude that the gas-dynamical model of the black hole mass is not correct.

It is at this point that modified gravity can step in. Let us consider my modified gravity theory, MOG, and the exact solution of a MOG rotating black hole derived from the field equations. The solution contains the mass M, the Kerr angular momentum parameter a, and the MOG parameter α. These parameters have to be measured for a given rotating black hole. The prediction of the size of the angular diameter of the shadow and the photon sphere is, in this gravity theory, enhanced by the parameter α. When α is equal to 1.13, and it is assumed that the mass of M87* is 3.5 $\times 10^9$ solar masses (obtained from the gas-dynamical model), then MOG predicts the measured angular diameter of 42 microarcseconds. This value of α is chosen because it fits the EHT data.[13]

Recall that when α equals zero, MOG reduces to the predictions of general relativity. Also recall that α determines the strengthening of Newton's gravitational constant as part of the theory's field equations. It is clear that

13. J.W. Moffat and V.T. Toth, "Masses and Shadows of the Black Holes Sagittarius A* and M87*in Modified Gravity," *Physical Review D*, **101**, 024014 (2020).

when MOG uses the lower value of the measured mass of the black hole obtained from the gas-dynamical model, it can then predict an angular diameter size for the shadow and the photon ring that is in agreement with the EHT data.

The difference between the predictions of general relativity and MOG for the angular diameter of the shadow and photon sphere is approximately a factor of 2. Although MOG uses the lower value of the M87* mass from the gas-dynamical model to match MOG's prediction to the EHT data, it is tempting to choose the higher mass value for the black hole because it is consistent with the prediction of general relativity. However, this ignores the fact that an independent measurement of the black hole mass is still an outstanding problem. To resolve this problem, more detailed investigations of the two mass determination models' assumptions and their consequences are needed, and more accurate data are also required. Hopefully, such research will lead to a resolution of the problem.

The EHT collaboration has claimed in their publications that from the measured angular diameter size of the combined shadow and photon ring and surrounding photon sphere, the mass of M87* is 6.5×10^9 solar masses with an error of 15 percent. Recall that this would be consistent with the prediction of general relativity. A publication by Samuel Gralla, Daniel Holz, and Robert Wald at the University of Chicago and the University of Arizona has investigated the role of the observed size of the dark central area and the photon ring light emission region, and the EHT image reconstruction width of these regions.[14] Other authors have claimed that a robust feature of light emission models is that the observed emission will peak near the photon ring. But, Gralla, Holz, and Wald claim that this conclusion is not correct. The peak of emission, they say, is so narrow that it never makes a significant contribution to the observed flux of radiation. Indeed, the photon ring cannot be relevant for the EHT observations. The problem, ultimately, is that the model fitting of the emission region and shadow is blurred, and therefore it is difficult for the EHT collaboration to

---

14. S. E. Gralla, D.E. Holz, and R.M. Wald, "Black Hole Shadows, Photon Rings, and Lensing Rings," *Physical Review D*, **100**, 024018 (2019).

resolve it, because the blurring washes out the sharp lensing ring features. We conclude that the validity of the EHT mass measurement is dependent on detailed physical assumptions and their correctness. These assumptions play an important role in the simulated images used to fit the observations and the mass of M87*.

In the case of Sagittarius A*, the stellar-dynamical model produces what are believed to be very accurate determinations of the mass of the black hole Sgr A*. This is because Sgr A* is a thousand times closer in distance, and the orbiting stars near the black hole can be well re-solved by infrared optical measurements. Moreover, the gas-dynamical model calculations for Sgr A* are reasonably consistent with the stellar-dynamical model calculations. Notwithstanding this, the EHT project has not yet succeeded in obtaining a value for the angular diameter of the shadow and the photon emission region of Sgr A* because of the problems of observing it within our galaxy, and the rapid changes in ra-diation from the black hole system. The size of the shadow and photon emission region angular diameter is about the same size as for M87*, but it is more difficult to observe.

It is interesting to consider that the measurement of the perihelion pre-cession of Mercury, performed over 300 years, leads to a total precession of 5600 arcseconds per century. The prediction by Einstein's gravity theory of the perihelion precession of Mercury deviates from Newtonian gravity by only 42.98 arcseconds per century. It is remarkable that this tiny predicted deviation is less than one percent, and yet it was a major reason for the eventual acceptance of Einstein's modification of Newtonian gravity.

The prediction of Einstein's gravity that a black hole will have a shadow with a specific angular size is a much bigger effect than the prediction of the perihelion advance of Mercury. However, with the variation in the estimates of the mass of M87* of as much as a factor of 2, the ability to dis-criminate between general relativity and modified gravity theory is cast in a different light.

In contrast to the solar system measurements, such as the perihelion precession of Mercury, which have taken hundreds of years to collect, the data for the supermassive black hole M87* observations by the EHT project are new, and provide a fresh way of evaluating gravity theories. In particular, the gravitational field of the supermassive black hole is much stronger than the weak gravitational field in the solar system.

With the ongoing detection of gravitational waves by the LIGO/Virgo project, and the EHT observational data for supermassive black holes, we have embarked upon an exciting time of both testing gravitational theory and increasing our understanding of black holes. The increased sensitivity of the LIGO/Virgo observatories since the beginning of April 2019, when "observation run 3" began, will allow for the detection of more black hole and neutron star mergers. The LIGO collaboration expects that an event will be detected every week or two, and this increased amount of data will shed much more light on gravitational waves and the nature of the compact objects that collide to produce these waves. The collaboration plans to increase the sensitivity of the LIGO/Virgo observatories and include the observations from the KAGRA observatory in Japan. Then, the mergers of black holes, neutron stars, and black hole–neutron star mergers will be detected as often as one per day.

Compared to the LIGO/Virgo collaboration, the EHT data, although dramatic, are from one source (until the measurements of Sgr A* and possibly other supermassive black holes can be achieved). The gravitational wave data are from many sources, producing a more encompassing picture of black holes and strong gravity. However, the combination of all the data from both projects has opened the door to a whole new era of observational astrophysics and gravitation, which will enable us to learn much more about black holes.

In time, we expect that the observations by the EHT collaboration, using the planet-size VLBI techniques, will produce more direct observational evidence and detailed information about black holes. In particular, the shadow cast by the supermassive black holes at the center of our galaxy and the galaxy M87 will, with sufficient accuracy, distinguish general relativity from modified gravity theories such as MOG. Moreover, a direct observation of the gaseous accretion disk surrounding these supermassive black holes can possibly provide unequivocal evidence for the existence of the event horizon and other significant properties of black holes.

Up until now, the evidence for the existence of black hole event horizons has been based on astrophysical calculations simulating the behavior of these accretion disks with and without possible event horizons. Although these calculations provide support for the existence of event horizons, they have loopholes in their arguments and do not provide completely convincing evidence, as yet, of the reality of event horizons. With the EHT experiments, we are entering a new and exciting phase of black hole astronomy that promises to reveal, finally, the true nature of these awesome physical objects.

FURTHER READING

Al-Khalili, Jim. *Black Holes, Wormholes, and Time Machines*. Boca Raton, FL: CRC Press, Taylor & Francis Group, 2012.

Bartusiak, Marcia. *Black Hole: How an Idea Abandoned by Newtonians, Hated by Einstein, and Gambled on by Hawking Became Loved*. New Haven, CT: Yale University Press, 2015.

Begelman, Mitchell and Martin Rees. *Gravity's Fatal Attraction: Black Holes in the Universe*. New York, NY: Cambridge University Press, 1995.

Eddington, Sir Arthur. *The Internal Constitution of the Stars*. Cambridge, UK: Cambridge University Press, 1926.

Eddington, Sir Arthur. *The Mathematical Theory of Relativity*. Cambridge, UK: Cambridge University Press, 1923.

Fock, Vladimir and N. Kemmer, translator. *Theory of Space, Time and Gravitation*. New York, NY: Pergamon Press, 1959.

Hawking, Stephen. *Black Holes and Baby Universes and Other Essays*. New York, NY: Bantam Books, 1993.

Hawking, Stephen and Roger Penrose. *The Nature of Space and Time*. Princeton, NJ: Princeton University Press, 1996.

Kennefick, Daniel. *Traveling at the Speed of Thought*. Princeton, NJ: Princeton University Press, 2007.

Levin, Janna. *Black Hole Blues and Other Songs from Outer Space*. New York, NY: Alfred A. Knopf, 2016.

Moffat, John W. *Cracking the Particle Code of the Universe: The Hunt for the Higgs Boson*. New York: Oxford University Press, 2014.

Moffat, John W. *Reinventing Gravity: A Physicist Goes Beyond Einstein*. New York, NY: Smithsonian Books, Harper Collins in New York, Thomas Allen in Toronto, 2008.

Schilling, Govert. *Ripples in Spacetime: Einstein, Gravitational Waves, and the Future of Astronomy*. Cambridge, MA & London, UK: The Belknap Press of Harvard University Press, 2017.

Schutz, Bernard. *Gravity from the Ground Up: An Introductory Guide to Gravity and General Relativity*. Cambridge, UK: Cambridge University Press, 2003.

Susskind, Leonard. *The Black Hole War: My Battle with Stephen Hawking to Make the World Safe for Quantum Mechanics*. New York, NY: Back Bay Books, Little, Brown and Company, 2008.

Thorne, Kip. *Black Holes & Time Warps: Einstein's Outrageous Legacy*. New York, NY: W.W. Norton, 1994.

Thorne, Kip. *The Science of Interstellar*. New York, NY: W.W. Norton, 2014.

Tolman, Richard C. *Relativity, Thermodynamics and Cosmology*. Oxford, UK: Clarendon Press, 1934.

Tyson, Neil deGrasse. *Death by Black Hole and Other Cosmic Quandaries*. New York, NY: W.W. Norton, 2007.

Figures are indicated by *f* following the page number.

For the benefit of digital users, indexed terms that span two pages (e.g., 52–53) may, on occasion, appear on only one of those pages.